Jane Roberts

The Whitehall Nightmare

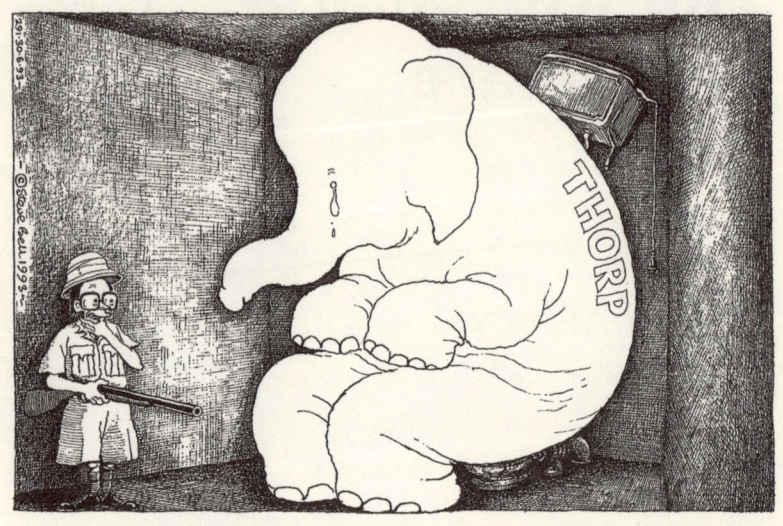

UK Environment Minister John Gummer ponders the THORP dilemma.
(Steve Bell in the *Guardian*, 6 June 1993)

THORP:
The Whitehall Nightmare

Crispin Aubrey

Jon Carpenter
OXFORD

Also by Crispin Aubrey:

Who's Watching You? (Penguin, 1991)
Meltdown (Collins and Brown, 1991)

ACKNOWLEDGEMENTS
The author is grateful to Steve Bell for permission to reproduce the cartoon on page ii, and to British Nuclear Fuels Ltd for permission to reproduce the photographs on pages 15, 16, 17, 59 and 75.

First published 1993 by
Jon Carpenter Publishing
PO Box 129, Oxford OX1 4PH

ISBN 1 897766 07 6

The publisher is grateful to *The Ecologist* for assistance with the publication of this book

Designed and typeset by Sarah Tyzack
Printed and bound by Biddles Ltd., Guildford, Surrey

Contents

The Sellafield plant in Cumbria

Introduction

SELLAFIELD has never been a popular British institution. When I enquired about trains to the area at my local rail station, the assistant said, almost without thinking: "You'll be coming back a different colour then." The name has become synonymous with pollution, cancer, accidents, nuclear waste and the unseen hazards of radiation.

When you visit the site it's a different story. There are real people working there who have only one head and who, despite the accepted risks, actually value their jobs. They cannot understand why they have been made the pariahs of British industry.

At the centre of these two opposing views has now been set a new controversy. THORP, the Thermal Oxide Reprocessing Plant, was hoped to be the venture that would rehabilitate Sellafield into the national and international community. In fact, it has done exactly the opposite. It has reawoken fears that Britain will become a dumping ground for radioactive waste that other countries don't want.

Are those fears justified, and the workers on the reprocessing lines simply clinging to a blind faith that their industry must survive? Or is THORP, as British Nuclear Fuels claims, a wonder of the world that will open up a new chapter in British technological advance?

THORP is not just about opposing visions, however. There are wider issues involved. The development raises fundamental questions which extend far beyond the boundaries of Sellafield. These are questions about the handling of the aftermath of nuclear power, the best way to dispose of nuclear waste, the uses and risks of plutonium, the health effects of radiation and the dangers of an international trade in radioactive materials.

The British government, which ultimately holds the purse strings for Sellafield, has now been thrust into the middle of this debate. It must decide whether THORP will open or not. Will that decision be made solely on the grounds of narrow economics and the fact that the building is already there and ready to operate? Or will the White-hall bureaucracy decide to take a step back—and reassess a project whose justification now looks very different to how it appeared when THORP was planned 20 years ago?

As the future of THORP still hangs in the balance, this book is an attempt to summarise the debate. It is not a full technical and scientific assessment, and readers looking for further detail will have to pursue some of the material in the references. But I hope it presents the arguments in a readable form that will be useful to those who have not yet made up their minds, or who want a handy guide to the major issues.

Whether THORP operates or not is a decision which could affect all our lives. It's important that we get it right.

1 The Background

THE Sellafield site in Cumbria lies on the coast of the Irish Sea, 25 miles north of Barrow-in-Furness and 20 miles west of Lake Windermere. It began life in 1939 as a factory manufacturing explosives. For a while towards the end of the War there was a chance that it would be used by the Courtaulds company to produce rayon material for clothing in the peacetime consumer boom. This plan was abandoned in favour of a more urgent role—in the post-war race to make an atomic bomb.

In 1947 work began on two simple air-cooled reactors for the production of plutonium to be used in Britain's first nuclear weapons. They were known as the "Windscale piles" after the name then given to the site, and their tall brick chimneys still dominate the landscape today. Their operation was crude by modern standards: metal rods of uranium fuel were pushed into channels in a giant graphite honeycomb, and, once irradiated by the nuclear reaction, ejected from the back. Hot gases from the reaction went up the 400 feet high brick chimneys, supposedly filtered before they emerged into the atmosphere.

Around the Windscale piles a network of other buildings was erected to complete the process of producing the explosive heart of a nuclear weapon. A chemical separation plant was built to extract plutonium from the irradiated fuel rods—the very first "reprocessing" line. A plutonium finishing plant produced the finished metal for the bombs, and a storage plant was commissioned to store the highly radioactive waste left over at the end of the process. In 1954 the completed complex was handed over to the freshly created United Kingdom Atomic Energy Authority (UKAEA).

Eventually, a new design of more sophisticated reactors than the "piles" was developed to reclaim plutonium for the expanding military demand. Known as Magnoxes, after the magnesium alloy used for the fuel canisters, the first examples (four small reactors in a row with large cooling towers) were built just across the Calder river from Windscale. The Calder Hall "power station" was opened by the Queen in 1956 with a champagne-cracking ceremony to mark the fact that the "miracle" of nuclear power had produced its first electricity. No mention was made in her speech of the main purpose of the reactors—producing plutonium for nuclear weapons.

Within a year the first shadow had been cast across the Windscale site. In October 1957, an attempt to disperse a dangerous build-up of energy inside one of the twin plutonium piles ended in disaster. The reactor caught fire, sending a cloud of radioactive dust up the chimney and out across the Cumbrian countryside. As a precaution against contamination, two million litres of milk from surrounding farms had to be tipped down the drains. The radioactive cloud eventually spread south right across Britain and even into Europe. It only emerged afterwards that, due to the inefficiency of the chimneys' filters, particles of burnt fuel, containing radioactive plutonium, had been routinely falling on the surrounding area for several years before the fire.

The disaster was over-shadowed, however, by the burgeoning development of nuclear power stations around the country with their promise of cheap, safe, clean electricity to power the post-war boom. Starting in 1962, nine civilian Magnox stations were opened at a cost of over £1 billion. These were among the first nuclear power stations in the world, and two were even exported—to Italy and Japan. To reprocess their "spent" (used) fuel, now with the intention of reclaiming both plutonium and uranium, the UKAEA started up a second, larger reprocessing line at Windscale, known as building B205. It began accepting Magnox fuel in 1964.

Meanwhile, work had started on a second generation of British-built nuclear stations. These had a larger capacity, operated at a higher temperature, and, although also cooled by carbon dioxide gas, used a new type of uranium fuel packed into stainless steel tubes. A prototype "advanced gas-cooled reactor" (AGR) was built at Windscale and started operation in 1962. Although now shut down, its large metallic spherical housing is still a landmark at the site.

Construction of the AGR series of reactors turned out to be a financial and technological nightmare. Although seven power stations of this design were eventually built, the problems experienced, especially at the Dungeness site in Kent, have led the reactor series to be dubbed "the largest loss making civil project ever undertaken in the UK".[1]

Even so, for the operators of Windscale it was the chance to provide another service to the nuclear industry. A new reprocessing system was investigated to deal with the different demands of the "oxide" (enriched uranium dioxide) fuel from AGRs. Oxide fuel was also being used in the majority of designs for nuclear reactors now being built in other countries around the world, most of which employed pressurised water rather than gas as a coolant. Here was the opportunity for what appeared at the time to be a valuable international trade in reclaiming uranium and plutonium from other countries' nuclear fuel—and dealing with the resulting wastes. For British Nuclear Fuels Limited (BNFL), the state-owned company which took over the Windscale site in 1971, this was the beginning of THORP.

DURING 1974, together with its French counterpart COGEMA (Compagnie Générale des Matières Nucléaires), BNFL began private negotiations with a consortium of 11 Japanese electricity utilities to reprocess their spent nuclear fuel. They were to be the first major overseas customers for the centrepiece of an ambitious expansion plan at Windscale—an oxide reprocessing plant which would handle up to 2000 tonnes of spent fuel a year and become a magnet for the world's nuclear operators. The following year, as more details emerged, the *Daily Mirror* warned in a splash front page headline that the plan would turn Britain into the "World's Nuclear Dustbin". The controversy had begun.

In fact, this was the second attempt to reprocess spent oxide fuel at Sellafield. The first had ended in disaster in 1973, when a serious accident (of which more later) forced complete closure of the pilot plant. By this time, however, deliveries of fuel had already been agreed with overseas customers, pushing BNFL into trying again.

Despite this background, THORP itself might still have been built without any lengthy public debate. Although there was mounting opposition from pressure groups, including a petition of 27,000 names, the planning application, submitted in June 1976, was generally accepted by the local councils, who anticipated at least 1,000 new jobs. The decision on whether to call a public inquiry was left to the then Labour Environment Secretary Peter Shore, who was clearly torn between the pressures to earn foreign currency and the growing concerns of Cabinet colleagues like Tony Benn, who had called for a national debate about nuclear power.

One incident may well have tipped the balance. On 10 October 1976, a concrete storage silo at Windscale was found to be leaking 200 gallons a day of radioactively contaminated cooling water. No public announcement had been made for 12 days. Soon after this, Peter Shore announced that there would, after all, be a full inquiry into THORP.

The Windscale Inquiry, chaired by Mr Justice Parker, heard evidence and submissions in Whitehaven Civic Hall during 100 days between June and November 1977. It was the first full-scale formal confrontation between the nuclear industry and the anti-nuclear movement, and set the pattern for later, even more lengthy debates at Sizewell and Hinkley Point.

British Nuclear Fuels were opposed at the inquiry by a number of environmental and other organisations, from the Town and Country Planning Association to Friends of the Earth to the British Council of Churches, as well as umbrella groups of smaller "green" objectors. The company was publicly supported by the local Copeland District and Cumbria County Councils, by the English and Scottish electricity generating boards, and tacitly by the various regulatory bodies, who raised no over-riding objections to the plan. A sign of the importance accorded the inquiry was the coverage given in *New Scientist* magazine, which ran full page articles on the evidence every week.[2]

It is generally accepted that Friends of the Earth (FoE) put up an impressive and authoritative objection to THORP. Against BNFL's argument that the plant was the best way to handle spent fuel, and would produce useful by-products in reusable uranium and plutonium, FoE presented a case which has a telling resonance 16 years later.

FoE's case characterised THORP as the centre of a new departure in energy policy which would lead inevitably to a plutonium-based economy. This had two strands, one extending towards the prolifer-ation of nuclear weapons, the other towards a new generation of untried, potentially dangerous "fast" reactors (a subject specifically excluded from consideration at the inquiry). It also warned that THORP would be "unnecessarily expensive and commercially risky". Instead of allowing the plant to be built, FoE advocated the continued storage of spent fuel at Windscale for ten years, during which time it would become apparent whether reprocessing—or some type of long term storage—was the best option.

Many other issues were touched on at the Windscale Inquiry, including the controversial area of the plant's emissions of radioactive material into the air and sea. I shall return to these later, as well as looking again, especially with the passage of time, at the arguments presented by Friends of the Earth.

Mr Justice Parker was clearly not impressed with the FoE case. Within much less than 100 days he had produced a report which, in less than 100 pages, gave a barely qualified endorsement to the BNFL plans. FoE felt severely betrayed, and the *New Scientist* accused Parker of "misrepresenting" the opponents' views.

For the politicians, on the other hand, there were few doubts. Environment Secretary Peter Shore said that he found the report "persuasive and broadly acceptable". With the exception of *The Observer*, which has consistently opposed THORP, there was little dissent in the national media. After a debate in the House of Commons the project finally received government approval in May 1978. Within weeks the Japanese contract which would allow the construction work to start had been signed. One of the world's largest nuclear reprocessing plants, and one of the biggest civil engineering projects ever seen in the North West, was off the starting blocks.

2 The Dream of Endless Energy

FIFTEEN years after the Windscale Inquiry participants tussled over the issues, it might seem extraordinary that THORP is still a centre of controversy. But almost a year since the building was completed it has still not been opened, and the opposition to it now is even stronger than it was when that almost forgotten dinosaur, a Labour government, held power at the end of the 1970s.

To understand why, it's necessary to look at the role THORP had been expected to play within the nuclear industry, at its basic justification, and the ways in which its apparently unassailable pillars have been dramatically shaken.

When THORP was first envisaged, during the early 1970s, the world's energy situation was in a very different state. In 1973, the volatile political situation in the Middle East exploded with a dramatic decision by a consortium of oil-producing Arab states to severely curtail their supplies of fuel to the Western countries as punishment for their support for Israel. Oil prices immediately soared, a 50 mph speed limit was introduced in Britain and there was a three day working week over the following winter. So serious was the crisis that petrol ration books were printed for the first time since the Suez war.

Against this background, radical ideas were sought right across the energy spectrum as economists and politicians rethought their accepted dependency on the black gold. For the nuclear industry, then beginning to spread its wings, it was a golden opportunity. Quite apart from its claimed advantages as a pollution-free source of power, nuclear energy had the benefit that its original fuel source—uranium—was mostly located in countries with close ties to the energy guzzling West. The main producers of uranium were then (and still are) Australia, Canada, South Africa, Namibia, France and the United States, as well as the African states of Niger and Gabon.

So optimistic was the expectation of an increase in the use of nuclear power that in 1975 the UKAEA projected that by the end of the century there would be 104 Gigawatts of nuclear capacity in Britain. One Gigawatt (GW) is the electrical output of an average nuclear power station. By the year 2030, in a fast expanding economy, the Authority expected there to be 426 GW of capacity. At the time there were just 5 GW of nuclear stations in commercial operation.

One consequence of these expansionist plans, however, which were paralleled in other nuclear states as well, was an immediate increase in the price of uranium, and a feverish rash of mining exploration. Between 1973 and 1976 the price of uranium quadrupled, from about $20 per lb to over $80.[3] At the Windscale Inquiry, a representative of the Uranium Institute warned that by 2010 the world supply of uranium would have run into such severe shortages that the whole future of nuclear power would be threatened.

One answer to this impending drying up in the supply of raw nuclear fuel was to do something which, on the surface at least, sounds environmentally attractive. The idea was to "recycle" left-over fuel by skimming off unwanted waste, cleaning it up and reclaiming some of its original material for re-use. This is the theory of reprocessing, a technology in which Britain, as one of the first

THORP's Twenty Year Progress

1969   First uranium oxide reprocessed at Sellafield site in "Head End" plant.

1973   Head End plant shut down after accident.

1974   Negotiations begin with Japanese power companies. Full scale oxide reprocessing plant (THORP) planned.

1976   Planning application submitted to Copeland Borough Council.

1977   Windscale Public Inquiry into THORP opens at Whitehaven.

1978   Inquiry report gives approval for THORP. Ratified by parliament. Japanese contract signed.

1983   THORP construction begins.

1992   Main construction work completed. Public consultation over new discharge authorisations.

1993   Further public consultation over justification for THORP. BNFL receives permission for test using radioactive uranium.

countries to develop nuclear power, and also one without its own indigenous source of uranium, was already a leader.

The industrial process involved in reprocessing is straightforward if not necessarily easy to perform. Nuclear fuel has to be removed from reactors after a given time because it has become too contaminated with radioactivity to remain efficient. It is then usually stored under water at the power station for a few months before being despatched in special containers to the reprocessing plant.

After further temporary storage in deep cooling ponds, to allow some of the shorter-lived radioactive "fission products" to decay, the fuel (long, thin metal rods in the case of oxide fuelled reactors) is then literally chopped up into more manageable pieces. These pieces are fed into a hot acid solution, which dissolves the fuel and enables the shredded metal fuel casings to be removed. A chemical separation process then allows the liquid, rather like a highly radioactive soup, to be divided into a series of different streams.

One of these streams produces uranium, which can then be reconstituted into fresh reactor fuel. Another produces plutonium, an element hardly found in nature but essentially created during the nuclear fission process. Finally, there is a mixture of different radioactive by-products from the process which have no useful after-life, but must be treated as hazardous wastes.

Alongside uranium, the recovery of plutonium was seen, at least at the time of the Windscale Inquiry, as the other main attraction of reprocessing. The manufacture of plutonium for nuclear weapons was, after all, the original purpose of the Windscale site. But in the late 1960s, as the nuclear researchers investigated new ways to exploit their armoury of radioactive materials, a further use had been found for this enticing element. This was as fuel in a completely novel design of nuclear reactor which, rather like the legendary perpetual motion machine, would reproduce its own basic necessities.

The first experimental "fast breeder" reactor was developed in the United States in the 1950s. Progress since then has been slow but the principle has remained the same. Using a core of plutonium fuel surrounded by a "blanket" of used uranium from ordinary nuclear reactors, the chain reaction literally "breeds" more plutonium from the uranium blanket. This plutonium can then be used as fresh fuel. The "fast" description refers to the fact that there is no moderator, as in normal reactors, to slow the reaction down.

The THORP plant under construction

Over a period of 20 years it is claimed that a fast reactor could breed enough additional plutonium to fuel another similar power station. It would also make extremely efficient use of potentially dwindling uranium supplies. The main drawbacks associated with fast reactors are the hazards of handling plutonium—a highly toxic substance with one of the longest radioactive half-lives—and the coolant employed to transfer heat from the nuclear reaction. This is usually liquid sodium, whose most dramatic feature is that if it comes into direct contact with water, the combination can be literally explosive.

Even in 1971, the Nobel Prizewinner Glenn Seaborg anticipated that so many fast breeder reactors would be built that there would be a shortage of plutonium to feed them. "We can foresee that the value of annual plutonium production in the US alone will exceed the value of the world's annual gold production around the year 2000," he wrote. "Some have surmised that plutonium could even replace gold as the international monetary standard..."[4]

In the real world, progress towards production of a commercial fast breeder reactor capable of generating economic electricity has all but ground to a halt. The French, who were the keenest to develop the technology, built the largest fast breeder in the world, known as Superphénix, at Creys-Malville on the Swiss border. The 1.2 GW reactor started up in 1985, but two years later suffered a serious accident when some of the liquid sodium coolant leaked out. At the time of writing, it is out of action following a further breakdown in 1990, and now cannot be started again without a full re-licensing process.

At the same time, a European research programme into fast breeders has been virtually abandoned. A decision to withdraw British government funding for the experimental fast reactor at Dounreay in the north of Scotland was taken in 1988 on the basis that it showed no sign of achieving commercial viability within an acceptable timescale. It is due to close in 1994. The German fast reactor at Kalkar, which started construction in 1973, never received an operating licence and was finally abandoned in 1991. The Italians have also withdrawn from fast reactor research.

Apart from Russia, where a 600 Megawatt demonstration fast reactor is still operating (and there are proposals for more), the only other country in the world to be pursuing this option is Japan. Even there, however, there have been delays and setbacks, as we shall see later when looking at the international ramifications of THORP.

The British nuclear industry still talks optimistically about an eventual programme of fast reactors. But the present situation must be set against the 1975 UKAEA projection, which expected there to be 33 GW of fast reactors in operation in Britain by the year 2000. It is now clear that there will not even be one.

Whilst fast reactor technology has failed to live up to its expectations, there has been a similar disillusionment with the wider promise of nuclear power based on more conventional reactor systems. In the United States, for example, which contains about a third of the world's reactor capacity, safety concerns, especially following the 1979 Three Mile Island accident, led to such an increase in the cost of nuclear construction that it priced itself out of the market. Growing public concern, exacerbated by the 1986 Chernobyl nuclear disaster in the former Soviet Union, has forced a moratorium on nuclear construction in Switzerland, Austria, Italy, Greece, and, in all but name, in Germany. Only in a handful of countries, such as France, has nuclear power been able to maintain any sense of momentum, albeit it on the back of massive government subsidy.

At the same time, world energy demand, especially in the "developed" countries, has not expanded as fast as expected. Energy users have learned to cope with increased costs, especially those inflicted by the Middle East oil price hike, by being more efficient and using less.

The end result of all these developments is that the expansion of nuclear power has now reached an international plateau where new orders for nuclear reactors are almost as rare as a swallow in midwinter. In 1992, according to the *BP Statistical Review of World Energy*, world nuclear energy consumption stagnated for the first time since the 1960s.

As far as uranium is concerned, whose uncertain supply was one of the main planks originally used to justify THORP, the effect of nuclear stagnation has been dramatic. By the beginning of the 1990s, the price had fallen to just $10 per lb—*half* of what it was before the 1973 oil crisis sent expectations soaring. Production in uranium mines around the world continues to exceed demand by a heavy margin, stockpiles mount up and the extraction industry is in crisis.

But if the original justification for reprocessing as a means of keeping the nuclear industry going into the 21st century has been heavily undermined, then there is one other issue that has not gone

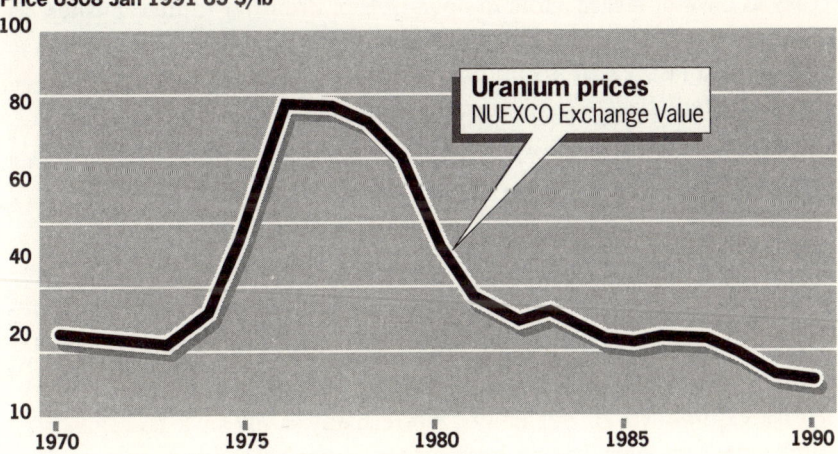

Uranium prices 1970-1990
Price U308 Jan 1991 US $/lb

Uranium prices
NUEXCO Exchange Value

away. This is the role which THORP would play in dealing with the inevitable aftermath of nuclear power generation.

Whatever the state of play in the uranium market, spent fuel produced by nuclear reactors will not disappear. Because it continues to be radioactive for thousands of years it will have to be dealt with safely—somehow, somewhere. British Nuclear Fuels argue that, apart from any other advantages from the operation of THORP, they are offering just such a waste management service.

Many critics, following the line of the sensational 1975 *Daily Mirror* headline, put it differently. They argue that nuclear power station operators, especially those from abroad, are only too happy to offload their nuclear rubbish into somebody else's backyard, even if it does cost a lot of money. It saves them the trouble of having to sort it out themselves. This has undoubtedly been a powerful motivation for THORP's customers, especially those fast running out of storage space of their own. But is reprocessing a satisfactory method of dealing with spent nuclear fuel, and what, quite apart from the arguments about recoverable resources, are its drawbacks?

THORP: The Inside Story

THORP is a massive structure at the heart of the Sellafield site—a long brown and cream building with red painted stairwells. Like a mammoth anonymous factory, it covers an area about the size of four Wembley football pitches. Its chimney is taller than St. Paul's Cathedral.

Inside, a rabbit warren of long strip-lit corridors lined by miles of pipework snake into the distance, requiring a map to negotiate. Red warning lights flash, and loudspeakers issue a steady "plick, plock", a sound that would turn into a siren if there was a real radiation alert. Occasionally you can peer through thick glass viewing windows into the plant's interior—bright stainless steel "caves" lit by an eerie orange light. Glove boxes enable operators to reach some controls, but because of the high levels of radiation, most operations are carried out remotely, usually through a computer programme. There are constant warning notices about the dangers.

THORP's operations begin with the arrival at one end of spent fuel in large white containers—five metre long cylinders for light water reactor fuel,

Storage pond in the Receipt and Storage area

squatter boxes for British AGR fuel. Staff cheerily refer to these containers as "coffins". AGR fuel is stored for at least three years before reprocessing, light water fuel for five years. Although some is being kept elsewhere on the Sellafield site, a fair quantity of fuel has already been removed from its boxes and shifted underwater into a large "cooling pond"—something like a giant radioactive swimming pool.

From this cooling pond, the fuel would be moved through into the "Head End Plant", where a heavy duty chopping device rather like a guillotine literally chops the steel fuel rods into smaller pieces. So much force is required that operators expect the chopping blades to have to be replaced every week, possibly every few days. This is a complicated operation carried out by remote control from outside the "cave", and a potential source of breakdowns. One employee described this shearing process as "the throttle of the whole business. If you have trouble there, the whole thing's in trouble."

The slices of fuel element are then fed into a dissolver, a large "tub" filled with nitric acid. To make sure that all the fuel has dissolved, the acid is superheated under pressure. After that, a large mesh "sieve" removes the

The Shear machine at the Head End

The Fuel Removal Pond at the Head End

The Dissolver Cell at the Head End

undissolved metal. This is incorporated into concrete blocks and treated as intermediate level waste. The remaining "liquor" then goes through a series of chemical processes, the end result of which is to separate out plutonium, uranium and the remaining liquid, which is classified as high level waste.

Although it sounds like a clean cut process, THORP in fact involves a large number of operations requiring extremely careful handling of hazardous materials. The separated plutonium liquid, for example, has to be kept in special tanks with tiny tubes so that there is never too large a concentration—with the possibility of it reaching criticality. Aerial discharges of radioactivity are given off through the chimney, mostly during the initial dissolving process. Liquid and solid wastes are produced at various stages of the operation.

There is no guarantee that THORP will run as smoothly as expected. The plant's chemistry has been tried out on a prototype model, and some mechanical parts have been tested. But no full scale trial has been possible. Past experience of Sellafield operations does not instill confidence.

3 The Waste Management Debate

ONE of the more intractable problems facing the nuclear industry since it first started has been how to deal with its waste products. The continuing radioactivity contained within its fuel and its buildings, even after they have finished their useful life generating electricity, presents quite different problems from those associated with other methods of energy production. There has also been a developing scientific controversy about the hazards associated with the very long periods over which some of these materials will remain radioactive.

The issue was highlighted in 1976, just before the Windscale Inquiry, by the 6th Report of the Royal Commission on Environmental Pollution (known as the "Flowers Report" after its Chairman, Sir Brian Flowers). In a now familiar injunction, Flowers warned that "there should be no commitment to a large programme of nuclear fission power until it has been demonstrated beyond reasonable doubt that a method exists to ensure the safe containment of long-lived radioactive waste for the indefinite future."

Whatever other criticisms Flowers had of nuclear waste management in Britain, however, he did not question the fundamental premise of reprocessing. It was equally assumed at that time within the British nuclear industry that it was the only logical first stage in the waste management process.

As already explained, reprocessing involves the dissolving of spent nuclear fuel in a powerful acid solution. But even after the plutonium and uranium streams have been extracted, and the recoverable materials stored, there are other streams of waste material which require careful but different treatment.

The dissolution process in fact increases the overall volume of radioactive waste to be dealt with by about 50 times. In addition, a whole range of materials—from plastic gloves to tools to pieces of equipment and even the buildings themselves—will become contaminated during reprocessing and must therefore also be treated as radioactive waste. The total increase is therefore about 189 times the original volume of the spent fuel.[5]

The industry divides radioactive waste into three broad categories. Low level waste, mainly clothing and other equipment contaminated during reprocessing, is the least radioactive, but is still packed inside special containers and isolated in a concrete storage bay. At present, this is carried out at Drigg, a few miles from the main Sellafield site. Intermediate level waste poses greater hazards, and must be tightly packaged and kept totally isolated from contact with human activity. This waste is currently stored in special vaults on the Sellafield site.

The most hazardous waste is high level. This is the concentrated liquor left over from reprocessing once the less radioactive streams have been removed. It contains over 90 per cent of the radioactivity contained in the original fuel rods, and generates intense heat as some of its elements steadily decay. At present it is stored in large stainless steel and concrete silos which must be constantly agitated to stop the liquid precipitating out. When cooler, it is gradually being "vitrified" into glass blocks ready for eventual disposal.

The vitrification plant has been plagued with breakdowns and technical problems since it opened in 1991, however, and has only operated at half capacity. This year, BNFL was prosecuted and fined for breaking safety rules at the plant. One possibility is that a second vitrification line will have to be built to deal with the backlog.

The growing quantities of lower level wastes being produced both at Sellafield and at nuclear power stations, meanwhile, mean that the Drigg site is now nearing saturation. In 1987, the nuclear industry's waste management company, NIREX, estimated that between then and 2030 there would be a further 1.5 million cubic metres of low level and 250,000 cubic metres of intermediate level radioactive wastes to be disposed of.[6] The largest source of both types of waste was the manufacture and reprocessing of fuel.

After a series of abortive investigations around Britain during the 1980s, public and political opposition forced NIREX to concentrate its search for a new waste disposal site on Sellafield itself. This is now the potential site for a deep underground "repository" spacious enough to take these large quantities of low and intermediate level waste. Its current design would involve the excavation of as much rock as was removed to construct the Channel Tunnel, and cost about £3 billion to build. However, geological investigations have so far failed to provide convincing evidence that the waste would be kept isolated from the surrounding environment for the long periods—effectively thousands of years—during which its residual radioactivity will continue to decay. Although it is planned eventually to build a separate long term resting place for the vitrified blocks of high level waste, a search for such a site has currently been shelved.

Despite the uncertainties about ultimate disposal of the wastes produced, British Nuclear Fuels still argues that the advantage of reprocessing is that it transforms the spent fuel into a more manageable commodity. It removes the plutonium, with its extremely long half-life, for separate treatment, and it condenses the most radioactive waste into a concentrated form. It is this high level waste which is portrayed reassuringly in BNFL publicity as equivalent in size to a row of red London buses.

In practice, reprocessing and the treatment of its wastes have turned out to be a much less straightforward activity than the above description suggests. Since the original Windscale fire, now accepted to have been second only to Chernobyl in terms of the quantities of radioactivity released into the environment, resulting in up to 100 fatal cancers, there have been numerous other incidents (more than at any other UK nuclear site), and some involving substantial leaks of radioactive materials.

In 1976, for example, a silo storing the exterior cladding from spent Magnox fuel was found to have been leaking for at least four years, releasing an estimated 50,000 Curies of radioactivity into the ground. During 1979, four separate serious incidents involved two fires, the leakage of up to 100,000 Curies of highly radioactive liquor from a disused building, and the spillage of contaminated solvent.

The deteriorating safety record at Windscale, at that time (the late 1970s) resulting in about 30 incidents per year, prompted the Health and Safety Executive (HSE) to carry out a special investigation.[7] The HSE report, noting that about a quarter of the incidents had "resulted in the exposure of workers to levels of radiation exceeding statutory limits", concluded that many of them "might have been prevented, or the risk of their occurrence significantly reduced, had formal arrangements been in existence for the regular updating and review of procedures, and compliance with them."

Over the next few years, however, untoward events continued to occur. In 1983, one of the most dramatic followed the discharge into the sea of a quantity of contaminated solvent used during reprocessing after it had been accidentally released from its original storage tank. No precautions were taken until contaminated debris was washed ashore, resulting in the closure of many local beaches for almost nine months. British Nuclear Fuels was fined £10,000, with £60,000 costs, for failing to "minimise the exposure of persons to radiation" and failing to keep adequate records. This was the first time that a criminal prosecution had been taken out against any part of the British nuclear industry.

Three years later the Health and Safety Executive, which is responsible for monitoring safety in the industry, returned to the fray with another report. This time it reported that, despite many improvements, the "condition of the plant seems to have been subordinated to the requirements of current production, is unsatisfactory and demands planned new investment to enable it to perform for a further ten years and beyond without unnecessary hazard to workers, and in the extreme, to the public."

Studies of current Sellafield workers have in fact not revealed any dramatic evidence of general ill health. But since 1982, when a special compensation scheme was introduced, the company has paid out over £500,000 to the families of workers who have died from cancer after working at the plant. BNFL has never accepted liability for these deaths.

As far as the general public is concerned, the major issue has been the effect of routine radioactive discharges from the site, which are 20 times larger than from all the country's nuclear power stations put together. In the mid-1980s Sellafield accounted for as much as 90% of the UK public's dose from radioactive wastes.[8] Most concern centres on discharges into the Irish Sea, which are channelled through a two kilometre long pipeline. Radioactivity is washed out through a daily circulation of about two million gallons of sea water.

When the discharges first started in the 1950s they were said to have been deliberately maintained at "fairly substantial" levels in order to study their effects as part of a scientific experiment.[9] But as

Right: The covered pipeline carrying Sellafield's radioactive discharge into the Irish Sea

activity on the site grew, and more fuel was delivered for reprocessing, the level of discharges increased still further. Between 1970
and 1975, the amount released from the pipeline of one single
radioactive element, caesium, rose from 31,170 curies to 141,377,
mainly as a result of increased corrosion in spent fuel. The total
radioactivity released in 1975 was the highest ever—250,000 Curies.

There are various "pathways" through which radioactivity
released in this way could return to the human environment. These
include through eating fish or seafood, through external radiation
from the beach, or through ingesting particles blown inland. There
are also limits set on the discharges by regulating authorities, at
present Her Majesty's Inspectorate of Pollution (HMIP) and the
Ministry of Agriculture, Fisheries and Food (MAFF), based on
assumptions about the effects. However, this is an area where
official bodies and environmentalists rarely agree.

Particular concern has centred on those components of the
Sellafield waste cocktail which emit alpha radiation, such as plutonium and americium. These elements are both extremely long-lived
and toxic if consumed, as well as carrying the risk associated with
all radioactive elements of encouraging the development of cancer.
One theory is that they become locked into the sediments in the sea
bed and are then churned up, in more concentrated form, and
deposited along the coast. Readings taken along the creek of the
nearby Ravenglass Estuary have shown large variations in the
levels of alpha radiation, suggesting that official monitoring, taken
at greater distances, may miss out hot spots.

Since the peak of the mid-1970s, BNFL has been forced to limit
the radioactivity in the discharges. Two new clean-up plants,
including the Site Ion-Exchange Effluent Plant (SIXEP), opened in
1985, have reduced the levels. Even so, during the 1980s, keen fish
eaters along the Cumbrian coast were still receiving over half their
officially allowable annual radiation dose just from consuming
seafood.

It is also instructive to compare the discharge levels from Sellafield
with those from other reprocessing plants. At Cap de la Hague, the
French reprocessing plant near Cherbourg which carries out similar
operations, discharges into the sea during the 1970s and early 1980s
were consistently lower, as was the level of exposure for the
workforce. This is largely because treatment plant was installed to

limit radioactivity, particularly long-lived elements such as plutonium. Peter Taylor of the Political Ecology Research Group reports that doses to the public from Sellafield have been "1000 times what is considered reasonably achievable elsewhere, and up to 12 times the United States Environmental Protection Agency regulatory limit for such installations".[10]

There has been a parallel worry about what Sellafield's chimneys put out into the atmosphere. One example of this is the contamination of sheep, which can concentrate radioactive caesium inside their bodies by eating polluted grass. When this became an issue in the aftermath of the Chernobyl accident, it was found that the sheep also showed the clear "fingerprint" of Sellafield caesium.

If there has been concern about Sellafield's discharges, however, it was massively increased by the 1983 revelation in a Yorkshire Television documentary[11] that there was a marked excess of leukaemia among children living in the parish of Seascale, less than two miles from the site. An unusual and potentially fatal disease, leukaemia (cancer of the blood) has few clear causes, but one of them is excessive exposure to radiation. In Seascale there were ten times as many cases in children under 10 as would be expected from the national average.

Yorkshire TV's research was followed by a government inquiry and by a series of other investigations into the incidence of cancers, both round Sellafield and round other nuclear sites. Around Dounreay, for example, site of the experimental fast breeder reactor which also housed a small reprocessing plant, there was found to be a similar excess of leukaemia cases in young people.

COMARE, the Committee on Medical Aspects of Radiation and Health, which looked at both these "clusters" on behalf of the government, concluded that they supported the hypothesis that "some feature of the nuclear plants... leads to an increased risk of leukaemia in young people living in the vicinity".[12] COMARE's chairman, Professor Martin Bobrow, went further in a newspaper interview, saying that "the burden of proof now rests with the nuclear industry to prove that there is no connection".

Other studies have implicitly tried to demolish the underlying assumption that nuclear plants like Sellafield are killing children. It has been suggested, for example, that leukaemia clusters occur in places where there are no nuclear connections, or that a virus

Growing up in the shadow of Sellafield

brought in to the area by the original construction workers may be responsible. It's also argued that the officially recorded levels of radioactive discharges from reprocessing are simply not high enough to cause cancers in people living nearby.

The most disturbing explanation has been offered by Professor Martin Gardner, whose extensive study, published in 1990,[13] suggested that the radiation dose resulting in the leukaemias may have been passed on to the foetus through the father's sperm. Gardner and his team showed that there was an eight times higher risk than normal of conceiving a child with leukaemia among those Sellafield workers who had received a particularly high radiation dose. The blunt response from a British Nuclear Fuels spokesman at the time was to suggest that those workers who were worried about these findings should consider not having children.

At the end of 1992, two local families took BNFL to court over their claims that Sellafield had caused cancers. A mother whose ten-month old daughter had died from leukaemia and a woman who developed non-Hodgkin's lymphoma (a similar type of cancer) whilst working there, both claimed substantial damages. At the time of writing, a verdict is still awaited. Another 40 families are awaiting the outcome.

AGAINST this background, what difference would the operation of THORP make to the hazards of the Sellafield site? Firstly, it's important to realise that, despite its early promise, reprocessing has not become a routine activity for the international nuclear industry. Although six other countries have larger nuclear capacities than the UK, there is only one other operating plant, at Cap de la Hague in France, which has been developed to reprocess spent fuel on a comparable scale. Some nuclear nations have deliberately chosen not to pursue the reprocessing route, either for technical reasons, including the extent of waste materials produced, or because of concern about the uses to which the resulting plutonium might be put. I shall return to these issues, including alternative ways of managing spent fuel from nuclear reactors, in later chapters.

The origins of THORP itself are hardly auspicious. The very first reprocessing of oxide fuel at Sellafield was carried out in a disused building previously used for Magnox fuel. Although this plant had been operating since 1969 with a potential annual capacity of 400 tonnes, only 80 tonnes of spent fuel had in fact been treated before a serious accident occurred in September 1973.

The accident happened when the operators began to dissolve a fresh batch of fuel without realising that granules of radioactive material had been left in the bottom of the dissolving pan. These hot granules had both evaporated some residual liquid and heated up the base. Their action was therefore rather like pouring water on to a hot frying pan, only with more serious consequences. A steam explosion sent a burst of radioactive gas out into the building.

Although alarms went off to evacuate the area, these were initially ignored. In the end, staff dotted around the ten storey building had to be located individually and told to leave. Thirty-five men suffered serious skin and lung contamination. The building has been sealed up and closed ever since.

There is therefore understandable anxiety about the safe operation of a reprocessing plant like THORP. One particular hazard comes from the storage of liquid high level waste in steel tanks. A study by the Political Ecology Research Group (PERG) has traced the course of an impending disaster should the tanks lose their power supply, and the boiling liquid escape.[14] Just such a power failure occurred for four hours at the la Hague plant in France, and the PERG analysis has been taken seriously enough by the German regional authorities for them to refuse a licence for a proposed installation unless tank safety could be improved.

Even if it operates normally, THORP will certainly lead to an increase in the discharge of radioactive emissions into the local environment. According to an application made to the Pollution Inspectorate (HMIP) by British Nuclear Fuels, during the first year's operation of THORP, discharges into the sea of alpha radioactivity would almost triple, whilst those of beta radioactivity, such as caesium, would increase over four times. There would be even larger increases in the radioactivity discharged into the atmosphere.

British Nuclear Fuels argues that these levels reflect the development of a range of new plant, not just THORP. This includes the Enhanced Actinide Removal Plant (EARP), which will actually reduce radioactivity in the long term but will be used initially to "flush through" a backlog of stored effluent. The discharge of some radioactive materials would still reach about 75% of the maximum levels set by HMIP. The Pollution Inspectorate says these levels would "effectively protect human health, the safety of the food chain and the environment generally".

These assumptions are fundamentally challenged by (among others) Friends of the Earth, particularly in relation to what are called the "critical group"—those members of the public who, through their eating habits or occupation, are likely to receive the highest radiation dose from the plant's operation.[15] Around Sellafield, these are taken to be people who live along the coast and,

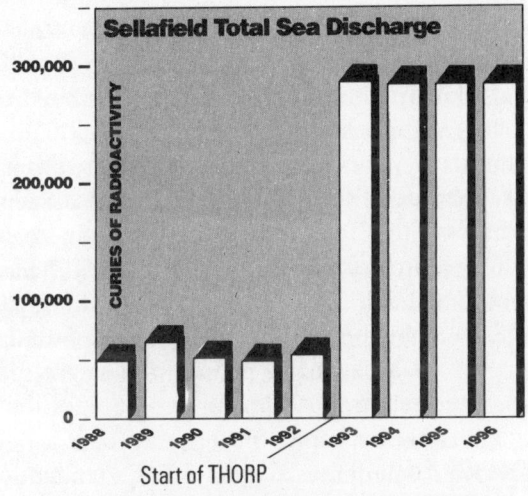

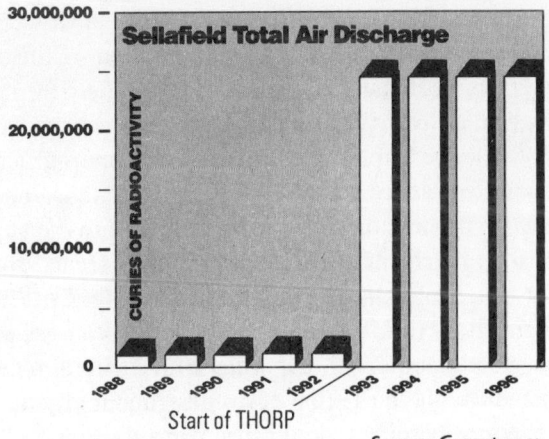

Source: Greenpeace

often through direct involvement in fishing, consume large quantities of fish or shellfish.

FoE says the authorities have failed to take full account of the latest international research into the risk of fatal cancer associated with a given dose of radiation. Following a progression which has been continuing for many years, this research shows that the risks have been previously underestimated. If taken into account by the British authorities, this would reduce the radiation dose considered acceptable to a member of the public by a third.

As far as the critical group is concerned, FoE says that the authorities have underestimated the likely effects of the two elements plutonium and americium (which concentrate in molluscs like winkles), particularly on the human gut. They have also assumed that local fish eaters will continue to maintain their consumption at the unusually low level recorded in the aftermath of the 1983 beach contamination incident. If consumption did return to pre-1983 levels, their radiation dose would rise dramatically.

Friends of the Earth have also carried out an extensive radiation monitoring programme of their own along the coast and estuaries around Sellafield. This has revealed unexpectedly high readings, especially along the muddy estuary banks. FoE argues that this not only presents a hazard to people who might spend time in these areas, it represents a baseline of exposure for the critical groups to which the new discharges must be added.

The net result of this analysis is FoE's contention that, if all these factors are taken into account, the annual radiation dose to the critical group of people living near Sellafield would exceed even the existing (and according to FoE over-generous) British limits on radiation doses to the public.

Again, it is worth contrasting the situation at Cap de la Hague in France. In 1984, the Department of the Environment set the aim of reducing Sellafield discharges so they were comparable with the French reprocessing plant. Meanwhile, however, the French have further reduced their levels to the point where, even if BNFL did achieve the 1984 target, its discharges would still be several times higher. That even greater reductions in discharges could be made is exemplified by the fact that when a German reprocessing plant was planned during the 1970s, it was designed to discharge (into a river) comparatively minute quantities of radioactivity.

Others have projected the proposed additional discharges from Sellafield into direct health effects. According to calculations by Dr. David Sumner, of Glasgow University's Department of Medicine, for Greenpeace, based on risk estimates from the International Commission on Radiological Protection, the expected toll for every year of discharge would be at least 60 deaths from cancer around the world.[16]

ONE particular aspect of the new discharges from THORP has caused an unusually high level of interest. This is the radioactive gas krypton. Although krypton is already released into the atmosphere from Sellafield's existing activities, THORP would increase the quantities ten-fold, from 100,000 to a million Terabecquerels each year. (Becquerels have replaced Curies as the international unit of radioactivity: 1 Terabecquerel = 27 Curies.) This would add 15% to the global inventory of krypton, most of which comes from the nuclear industry.

Concern about krypton centres both on its radiobiological effect and its influence on the climate. One theory, advanced at the 1977 Windscale Inquiry and since further investigated by German researchers,[17] is that krypton produces increased ionisation in the atmosphere and interferes with the electrical field. This could lead to both a reduction in precipitation and an increase in the total water vapour in the atmosphere. As well as the resulting problems of reduced rainfall, enhanced levels of water vapour could also add to the climatic changes and disturbances, such as violent storms, anticipated from global warming.

One of the specific recommendations of the Windscale Inquiry report was that BNFL should "devote effort to the development of plant for the safe removal and retention" of krypton, a condition accepted by the government. But after investigating the possibilities, the company withdrew from further research in 1982. It still argues that "safe and commercially viable krypton removal technology does not yet exist which is capable of application to a full-scale reprocessing plant".

Krypton removal plant is operated in the United States, but a Department of the Environment review has estimated the cost of introducing similar technology at THORP to be over £50 million, with between £2 and £3 million annual running costs. This is not considered to be justified in terms of the advantages obtained.

Since details of the proposed discharges for THORP were first published by BNFL in April 1992 there has been an unprecedented level of public interest. During a consultation period from November 1992 to January 1993, over 80,000 representations were received, the majority registering their opposition to the proposals.

There has also been adverse comment from the normally cautious members of COMARE, the committee which advises the government on environmental radiation issues and was responsible for investigating the leukaemia clusters. Having first complained that it was not furnished with enough information to accurately judge the health effects, especially on historical exposure to the local population, the committee then warned that, as far as Seascale was concerned, any increased discharges should be viewed with concern. It therefore couldn't rule out the possibility that there would be an increased risk to the health of local residents.

More fundamentally, COMARE reminded the government of the first principle of radiological protection, as suggested by the ICRP, the international advisory body. This was that "no practice involving exposures to radiation should be adopted unless it produces sufficient benefit to the exposed individuals or to society to offset the radiation detriment it causes". No estimate of detriment had been made available. The committee's comments were noticeably endorsed by the Department of Health.

THORP will also increase the quantities of solid radioactive waste which will have to be dealt with at Sellafield. This includes low and intermediate level waste resulting from the reprocessing of British spent fuel, whose intended ultimate destination is the (as yet unbuilt) NIREX repository, and high level waste to be vitrified awaiting an even more uncertain future. Altogether, the plant will spawn over 30 separate streams of different types of waste. Decommissioning of the giant building itself, which is expected to have a life of about 25 years, will also generate large amounts of waste.

It remains unclear how much of the waste produced as a result of foreign contracts will stay in this country. According to the small print of contracts signed with overseas customers since 1976 (the year after the Daily Mirror warned that Britain would become a nuclear dustbin), BNFL has the right to return all waste products to their country of origin. In practice this is not such an easy proposition, especially given the large volumes involved.

Since 1986, a scheme has therefore been devised by which THORP customers would have returned to them an equivalent quantity, in terms of radioactivity, of more compact high level waste. This would amount to about 14% of the total quantity of waste, leaving the remaining 76%—a mixture of low and intermediate level—to be disposed of in this country. Unsurprisingly, this arrangement, technically called "substitution", has aroused fresh concern about the dumping of other countries' waste in Britain, especially given the uncertain progress being made towards a long-term resting place.

The Radioactive Waste Management Advisory Committee (RWMAC), which advises the government on waste issues, has expressed particular disquiet, pointing out that substitution is fundamentally at variance with Britain's general policy on wastes. This is that developed countries should move towards self-sufficiency—in other words, clearing up their own rubbish. Once the Magnox fuel reprocessing programme is completed in the early years of the next century, the committee points out, foreign waste retained under substitution would account for up to 70% of the total Sellafield output. And if the NIREX underground repository is not opened on time, because of uncertainties about its geology, the extra waste from foreign customers would simply accumulate on the surface. RWMAC suggests that it might be more sensible instead to return intermediate level wastes, which are contaminated with unwanted plutonium.

The foreign customers themselves, such as Japan, are meanwhile clearly unprepared for the return of large quantities of wastes. Storage facilities are not even available for the high level waste, let alone the much larger quantities of other radioactive rubbish.

Perhaps the most telling comment has come again from the RWMAC. In its 1990 annual report, the committee concluded, having carried out a review of the waste management implications of reprocessing, that "there are no compelling waste management reasons to reprocess oxide fuel". The potential for dealing with spent fuel from nuclear reactors in a radically different way—the "dry storage" option—is discussed in Chapter 7.

4 Plutonium and Proliferation

I N the climactic moments of the TV drama *The Edge of Darkness*, the manic CIA agent sent to sabotage a secret British atomic plant staggers to his feet to address an audience of nuclear executives. Already dying from radiation sickness, his face pockmarked by sores, he waves two small dark-coloured blocks in the air and bangs them together. As an explosive flash lights up the room, the horrified onlookers rush for the door. Their terror is understandable. The two blocks, removed from an underground reprocessing plant, contain plutonium.

Plutonium's almost mythical aura originates from the fact that it is only produced in any quantity by the nuclear fission process. A dense grey metal, it occurs naturally in what are described as "infinitesimal" traces. It is literally a product of the nuclear age. But it has several other important qualities. If ingested in even minute amounts, it is both extremely poisonous and carcinogenic, and has one of the longest half-lives of any radioactive element. This means that it would take 24,000 years for half its radioactivity to decay. Most crucially, it is a vital component in the production of nuclear weapons.

Although already present in irradiated fuel as it leaves the reactor, reprocessing separates out the plutonium into a useable form. This is usually in the form of plutonium oxide powder. In order to ensure that the plutonium is suitably "fissile" for use in the manufacture of nuclear weapons, however, the fuel needs ideally to have been removed from the reactor after only a short period, usually a matter of months rather than the years it would normally remain there.

Regular removal and replacement of fuel rods in this way is generally not compatible with the efficient generation of electricity. Hence the existence in Britain of separate reactors, at Calder Hall in Cumbria and Chapelcross in Scotland, operated by BNFL on behalf of the Ministry of Defence, for the production of military plutonium.

The issue has not always been so clear cut. The first generation British Magnox reactors were specifically developed from a military design, and at least one of the resulting power stations, at Hinkley Point in Somerset, was deliberately adapted to ensure that it could produce military grade plutonium. Although its operators, then the Central Electricity Generating Board, vigorously denied that this facility had ever been used, there has been a lengthy controversy about whether plutonium resulting from the *civil* energy programme has eventually found its way into the *military* stockpile. In 1986, the then chairman of the CEGB, Lord Marshall, finally admitted that this had in fact happened in the past.

Even with more recent reactor models, such as the light water cooled design which is now the most popular around the world, there is still concern about the potential abuse of their resulting plutonium. Although this is generally impure compared with proper military plutonium, containing significant amounts of non-fissile isotopes, and therefore not ideal for nuclear weapons fabrication, it could still be used for that purpose.

The problems associated with trying to use reactor grade plutonium as weapons material include its volatility, leading to the possibility of pre-detonation, its higher radiation emissions, and the higher level of decay heat. However, none of these are insurmountable barriers. As long ago as the 1960s, the United States conducted a nuclear test using plutonium with a relatively high percentage of the unwanted plutonium 240 isotope. One US nuclear weapons expert has been quoted as saying that "it is likely that a nuclear explosive designer would choose to minimise the Pu-240 concentration—given the choice. However, an entirely credible national nuclear explosives capability could be constructed using reactor grade plutonium."[18]

The US Nuclear Regulatory Commission (NRC), which has responsibility for the country's civil nuclear programme, agrees. According to an NRC official, "so far as reactor grade plutonium is concerned, the fact is that it is possible to use this material for nuclear warheads at all levels of technical sophistication. In other words, countries less advanced than the major industrial powers but, nevertheless, possessing nuclear power programmes, can make very respectable weapons... Of course, when reactor grade plutonium is used there may be a penalty in performance that is considerable or insignificant, depending upon the weapon design. But whatever we might once have thought, we now know that even simple designs, albeit with some uncertainties in yield, can serve as effective, highly powerful weapons..."[19]

Stockpiles of "civil" plutonium have already accumulated from previous reprocessing activities. Up to the end of 1992, about 130 tonnes had been separated from spent reactor fuel around the world, 54 tonnes in Britain.[20] Most of that is still stored at Sellafield. Current world throughput is now adding to that figure at the rate of about 14 tonnes per year, a rate likely to increase by the end of the century as new plant comes on stream. THORP alone could produce over 60 tonnes of plutonium during its first ten years of operation.

By comparison, the quantity of plutonium needed to make a bomb can be measured in kilogrammes. The bomb which destroyed the Japanese city of Nagasaki at the end of the Second World War contained just six kilogrammes of plutonium. The possibility that the development of reprocessing plants could place nuclear weapons in the hands of new nations outside the established nuclear states was the main reason why the non-military use of the technology was abandoned by the United States in the 1970s.

Even in the aftermath of the Cold War, and the collapse of the Soviet Union, there are nations eager to acquire a bomb-making capacity. Apart from the officially acknowledged nuclear weapon states—the United States, France, China, Britain and the former Soviet Union—a number of other countries are known, or have admitted to having, weapons. These include India, Pakistan, South Africa and Israel. But a further lengthening list are either on the brink or may well already have developed a nuclear capacity, including Brazil, Argentina, Egypt, Iraq, Syria and Iran. Particular attention is currently focused on Northeast Asia, where both North and South Korea are known to be trying to develop a plutonium economy.

Some of these "unofficial" nuclear nations have developed their bomb-making potential through small scale reprocessing plants of their own. Others have acquired the technology through leaching out equipment and parts from the international civil nuclear industry, particularly in Europe. Examples include Iran, which has obtained material from a network of European companies, and Pakistan,

Right: The view from the beach: THORP under construction

which is believed to have obtained vital equipment from Germany.

All this has happened despite the existence of the Nuclear Non-Proliferation Treaty, an international agreement signed by 140 countries, which is supposed to limit the transfer of sensitive technology between nuclear and non-nuclear powers. However, there is nothing to stop the transfer of resources for the creation of nuclear power programmes to countries which have not signed the treaty. This is what happened in the case of India, which is assumed to have developed its first nuclear explosive using plutonium manufactured in a nuclear power plant supplied by Canada.

Many observers believe that the increased traffic in plutonium produced by the new generation of reprocessing plants, including THORP, will only exacerbate an international situation where proliferation safeguards are already in danger of falling apart. Two-thirds of the plant's first decade of plutonium production is due to be exported.

BNFL says that THORP has clean hands. The plant is not only subject to "safeguards" inspections operated by the International Atomic Energy Authority, which has responsibility for policing the potential diversion of nuclear materials from peaceful to military uses, but there will be a resident multi-national team of inspectors. Officials from Euratom, the European Community's safeguards agency, have also expressed their confidence in the system of audits and inventory controls aimed to ensure that all materials produced by THORP, especially plutonium, are accounted for.

Critics of the safeguards system say that, especially at facilities which handle materials in bulk form—as powders, liquids or gases—it is particularly difficult to keep track of them as they move through the factory. There has been an ongoing controversy in Britain about the alleged discrepancy between the amount of plutonium which should have been produced by historic reprocessing activities, and what appears in official records. In 1988, the shortfall was said to be between 1.5 and 3.1 tonnes.[21]

Even without these internal failures of the system, there is the ever present risk that someone may deliberately try to divert material for subversive purposes. This could happen through hijacking of a plane or ship carrying plutonium, or by smuggling small quantities out of a production plant. The "client" could be anybody from the IRA to an unscrupulous government determined to develop its nuclear "credibility". As we shall see in the next chapter, concern about proliferation of nuclear weapons remains a major part of international opposition to THORP.

WHILST plutonium has military uses, making it a material worthy of extreme caution, it also clearly has peaceful ones. We have already seen how the international plans to develop a new generation of plutonium-fed fast reactors have stumbled to a halt, if not totally collapsed. But if THORP's plutonium is unlikely, at least in the short term, to find a home in fast reactors, one other fuel route has now been developed. This is through a new type of reactor fuel known as mixed oxide, or MOX.

Fabrication of MOX fuel involves mixing a small percentage of plutonium oxide with the regular component of most nuclear reactor fuel, uranium oxide. It would be used mainly in pressurised water reactors (PWRs), the most commonly used design of light water reactor.

BNFL says that, once established, MOX fuel will make substantial inroads into the output of plutonium from THORP. The fuel could be loaded into existing PWRs without any modification, and would pose no safety problems. It would also be cheaper for a power company to use than ordinary uranium fuel. All these assertions are hotly contested.

MOX fuel assemblies can in fact only be loaded into reactors up to a level of 30% of the total fuel complement. At present, there are only a handful of small MOX fabrication plants around the world. These are in Germany, France, Belgium and Japan. Their annual maximum production capacity was estimated in 1990 to be 95 tonnes, for which about 4 tonnes of plutonium would be required. In that year, less than a third of the world reprocessing industry's output of plutonium was incorporated into reactor fuel.

There have also been problems with the operation of some MOX plants. At Hanau in Germany, for example, a 1991 breakdown, during which three workers inhaled plutonium dioxide powder, has led to an embarrassing backlog of plutonium waiting for delivery from the French reprocessing plant at Cap de la Hague. The German authorities have forbidden the import of plutonium unless it has an immediate use.

Plans for expanded MOX production capacity face further uncertainties. A second, much larger unit at Hanau is embroiled in a prolonged battle with the state government over operating licenses. Although a small pilot MOX plant is due to come on stream at Sellafield, plans for a full scale production facility, originally announced in 1986, are currently stalled by the protracted debate about THORP itself. In 1989, the plant was expected to cost £50 million to build. The figure given now is £300 million. On current form, MOX fabrication is likely to account for only a fraction of the plutonium produced by THORP and its French counterpart.

There is also a dispute about whether the use of MOX actually reduces the plutonium inventory. BNFL says that if a typical PWR was loaded with a third MOX fuel, it would contain about 500 kilogrammes less plutonium at the end of its useful life than at the beginning. Others assert that although some of the plutonium would be changed to other elements during the chemical reaction, the net effect would be to increase the inventory from about 17 to 20 kilos of plutonium per tonne of fuel.[22] One additional complication is that stored plutonium will degrade within a few years to a point where it is unsuitable for straightforward MOX fabrication.

There is little interest in the use of MOX fuel in Britain. It is not suitable for the uniquely British advanced gas-cooled reactors, and there is only one PWR, still under construction at Sizewell in Suffolk, in which it might be used. Nuclear Electric does not intend to use MOX at Sizewell.

As far as the economics are concerned, claims of savings on fuel costs for power station operators are treated with extreme scepticism. MOX is several times more expensive to manufacture than standard uranium oxide fuel, and claims that it could be 30% cheaper than an equivalent tonne of standard fuel, even assuming a zero cost for the plutonium input, are not accepted by the reactor users.

Who, it is argued, except those already committed (through reprocessing contracts) to the plutonium fuel cycle, would want to invest in a plant which is expensive to construct (because of security and radiation protection measures), will produce more costly fuel, and against a background of nuclear power stagnation? Most observers consider that, especially with a world surplus of uranium and prices at rock bottom, MOX is simply an expensive way of using unwanted plutonium. It also legitimates production of a material that has serious international security implications.

One of the heaviest assaults on the "plutonium economy" was made in 1992 by William Dircks, Deputy Director of the IAEA. Speaking in Japan, at the heart of the world's most ambitious nuclear industry, his message was simple: that the economic justification for use of recycled plutonium had been severely eroded, and that stockpiles of the material now presented a "major political and security problem".

Reminding his audience of the original expectation that plutonium would replace uranium as the primary reactor fuel, Dircks then rehearsed the familiar litany of reduced electricity demand, cancelled reactor orders, and the collapse in uranium prices. Whilst MOX fabrication plants were incapable of dealing with the growing quantities of plutonium being produced, yet further supplies could be released by the dismantling of nuclear weapons in the United States and the former USSR.

"Even if one disregards the fissile material from nuclear warheads," Dircks said, "the excess of isolated fissile plutonium from civilian nuclear programmes poses a major political and security problem worldwide. Although plutonium from power reactors tends to be impure—containing significant amounts of non-fissile isotopes—and not ideal for nuclear weapons fabrication, it can nevertheless be used for this purpose. Accordingly, it will have to be sorted under conditions of strict security and safeguards accountability."

Arguing that the best place for plutonium was in reactor fuel, Dircks asserted that "given the current availability of very low-cost uranium, there would appear to be little incentive to invest in additional facilities for the use of plutonium in commercial power generation. As we have already seen, the adverse economics of MOX fuel utilisation compared to using fresh, low-enriched uranium fuel will probably persist well into the next century."

Dircks concluded that "as a result of nuclear fuel reprocessing, and potentially as a result of nuclear weapons dismantling, in the foreseeable future the supply of plutonium will far exceed the industrial capacity to absorb plutonium into peaceful, commercial nuclear industrial activities." The most urgent issue was therefore to face up to the need to devise ways of storing plutonium safely and securely. One suggestion is that it could be stored centrally under the control or supervision of the IAEA.

As important as the message was the messenger. The IAEA not only has responsibility for proliferation issues but has a parallel role in advancing the technical development of peaceful nuclear power. That the second most senior official in this organisation should choose to give such an unambiguous warning about the dangers resulting from reprocessing was as clear a sign as any that concern had spread far beyond academic researchers and the environmental pressure groups.

5 The International Dimension

IN November 1992, the converted cargo ship *Akatsuki Maru* began a long voyage from the French port of Cherbourg. Its companion was a specially built escort vessel, the *Shikishima*, armed with two pairs of 35mm cannon, two 20mm machine guns and with two helicopters on its deck. The ship itself had a team of armed officers on board. During the two month journey back to Japan, the convoy was not planning to call at any ports or to take on fuel. The reason for these unusual security precautions resulted from its payload—the first large consignment of plutonium oxide to be separated at a French reprocessing plant and "returned to sender".

The *Akatsuki Maru*'s send off was dramatic. Watched by 150 journalists from around the world, the Greenpeace boat the *Moby Dick* tried to block its path but was arrested by the French authorities. On the quayside, a crowd of demonstrators was dispersed by 2000 riot police. Eventually escorted out of Cherbourg by French warships, the *Akatsuki Maru* was still pursued by another Greenpeace boat, the *Solo*. At one point the *Solo* was rammed by the convoy's escort vessel. Using a second pursuit boat, Greenpeace eventually gave up the chase near the Cape of Good Hope.

As the ship progressed around the world, there were further protests, this time from national bodies. The governments of South Africa, Malaysia, Portugal, Chile, Brazil, Argentina, the Philippines and Indonesia all requested that the ship keep out of their territorial waters. Before the voyage, the Asia Pacific Forum, representing 20 countries, called on Japan to halt the shipment. Australia, New Zealand, and some Caribbean countries also made strong protests. Japanese diplomats were reported to have made several sorties to try to allay the concern of countries along the route.

When it finally arrived in Japan, the ship, protected by 16 helicopters and 69 vessels from the Maritime Safety Agency, was greeted by further protests. One official is said to have described the exercise as "a public relations disaster". The cost of the voyage to the Japanese government was reported to be 6.3 billion yen, with a further 16.4 billion yen spent on building the *Shikishima.*

Clearly taken aback by the level of protest, the Japanese authorities said they might have to reconsider whether this was the best way to handle the regular series of plutonium shipments expected over succeeding years.

THE troubled voyage of the *Akatsuki Maru* was a salutary reminder of the international ramifications of nuclear reprocessing. It was also a foretaste of the arguments likely to ensue if THORP were allowed to start operation. The majority of British Nuclear Fuels' contracts for reprocessing at its new plant are with overseas customers, the largest single arrangement being with Japan. (See chart opposite showing THORP's order book for the first ten years). Current plans are for the first shipment of plutonium back to Japan to take place in 1995.

Concern about this traffic is not simply about its potential disruption by anti-nuclear groups, however, or about the ultimate use being made of its cargo, a subject to which I shall return. As the armed escort indicates, there are serious issues about its security both from accidents and from international terrorism. If any of the plutonium did escape from its containment, it would create a major radiological hazard.

Some plutonium is already transported around Britain and Europe by air, considered by the authorities to be the most efficient method in terms of cost and security. Armoured transporters are used to deliver it to the airport. But although this has aroused controversy, there has been even greater concern about longer, intercontinental flights.

In the 1980s, a flight carrying plutonium from the United States to Japan caused so much controversy that the Alaskan authorities refused permission for such aircraft to refuel. The United States subsequently introduced much stricter regulations for plutonium flights (under the so-called 1987 Murkowski amendment), including a stipulation that the shipment flasks should be tested in the equivalent of a worst case accident.

The tests included dropping the plutonium containers from a plane travelling at maximum cruising altitude and replicating a crash involving an actual cargo aircraft loaded with canisters. In the event, a real life crash of a large aircraft in the same year provided valuable data. Under the Nuclear Co-operation Agreement between the Japanese and United States governments, no plutonium is allowed by air into Japan unless these regulations are complied with.

At one time, British Nuclear Fuels had planned to fly its own output of plutonium back to Japan in Boeing 747 freighters taking off from Prestwick airport near Manchester. But the US-promoted regulations have proved so difficult to meet that this is not currently considered a viable possibility. Some observers doubt that a container satisfying the US requirements could ever be developed.

The alternative—long sea voyages round the horn of Africa and possibly also the tip of South America—are not without their hazards, however. Flasks used to transport the plutonium must be tested, for example, to withstand an 800°C fire lasting 30 minutes. But a 1992 report by a US engineering company pointed out that the average temperature and duration of seaboard fires was 1000°C and 24 hours. The report also noted that although IAEA standards require that a plutonium cask should be designed to survive immersion in water to a depth of 200 metres, much of a ship's voyage round the world would be through much deeper seas. All this, of course, is apart from the unpredictable hazards of maritime terrorism.

Even before the voyage of the *Akatsuki Maru*, a test shipment of

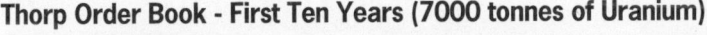

Thorp Order Book - First Ten Years (7000 tonnes of Uranium)

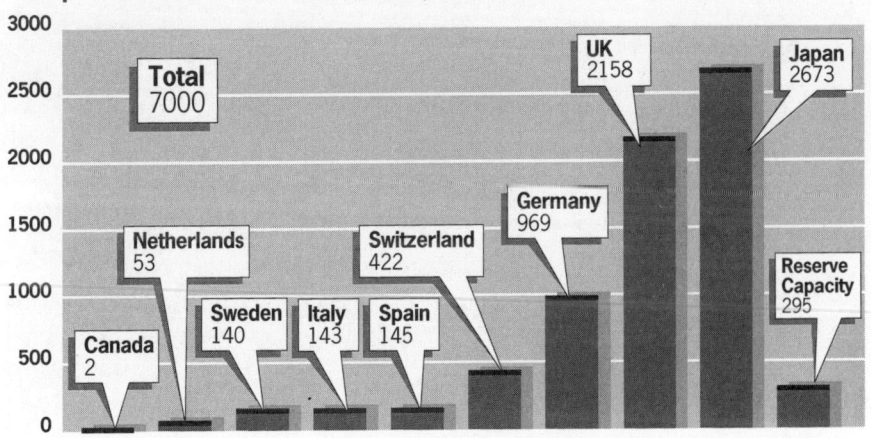

Total 7000

Canada 2
Netherlands 53
Sweden 140
Italy 143
Spain 145
Switzerland 422
Germany 969
UK 2158
Japan 2673
Reserve Capacity 295

Source: BNFL

Action against the *Akatsuki Maru* in Yokohama

Demonstration against the *Akatsuki Maru* in Tokai, Japan

250 kilogrammes of plutonium oxide was escorted from France to Japan in 1984. On that occasion military ships from France, Japan, the United States and Britain were all involved, as well as satellite surveillance. It is obvious that, apart from anything else, such elaborate precautions taken on a regular basis would add substantially to the costs of the whole plutonium fuel cycle.[23]

It is unclear at present how much return traffic in nuclear materials will be generated by THORP. This depends partly on what agreement is reached over the issue of "substitution", the system by which smaller quantities of highly radioactive waste are substituted for larger volumes of less radioactive materials. But the plutonium shipments will clearly be only the most hazardous outriders of a regular international trade in nuclear waste.

There is already concern about the incoming traffic, the shipments of spent fuel being delivered to THORP and now stored at Sellafield in "cooling ponds". Containing a cocktail of radioactive elements, this material, usually transported in massive cylindrical steel containers, also requires careful security measures.

For many years, a fleet of specially converted ships has operated from the port of Barrow, just south of Sellafield, bringing in spent fuel from Italy and Japan. These ships are fitted, for instance, with additional collision bulkheads, segregated cargo holds, special fire suppression systems and a "transponder" which enables information about damage and any leak of radioactivity to be remotely monitored should the vessel capsize. But the increase in traffic, especially from Europe, has meant that many other ports are now used. Spent fuel now arrives from Germany and Switzerland at Dover, for example, using a regular roll-on, roll-off freight ferry. Unlike Barrow, a port like this is not geared up to deal with radiation related accidents.[24] The European Parliament has called for a ban on the use of non-purpose built ships for such cargoes.

The safety of transporting spent fuel into THORP, and returning its output of by-products and waste, is only one aspect, however, of the project's international dimension. For the trans-national trade in nuclear materials also raises fundamental issues about the justification for reprocessing, at the centre of which is THORP's biggest customer, Japan.

IT is perhaps ironic that the country with the largest commercial involvement in THORP also happens to be just about the furthest away. But there have been powerful pressures from the Japanese political and industrial establishment to justify these global excursions.

Japan has few indigenous sources of energy. In order to power an expanding national economy, satisfy the demands of a growing population of over 120 million, and avoid expensive imports of fossil fuels, it has been determined to develop a self-sufficient energy policy. At the centre of that drive for energy independence has been nuclear power.

Since the oil price shocks of the 1970s, the Japanese nuclear industry, supported by both the government and the industrial ministry MITI, has steadily expanded to the point where it is now the fourth largest in the world. Mainly using a mixture of boiling water and pressurised water reactors, it has an operating capacity of over 33 GW and generates about 24% of the country's electricity.

There has been consistent pressure from the industry for continued expansion. In 1990 plans were announced to more than double the country's nuclear capacity by 2010 by building a further 40 reactors. But there has also been a growing anti-nuclear movement, strongly supported by women in particular, which has been equally determined to put the brakes on, most often by protests at new power station sites.

The most controversial aspect of the Japanese industry's plans has been its attempt to avoid dependence on imports of uranium by steadily moving towards a plutonium economy. There are already two small experimental fast breeder reactors in the country, and a pilot reprocessing plant. But there have been plans for some time to expand this capacity into a major network of fast breeder power stations. The initial stages of this were to build a larger fast breeder reactor at Monju, and a new, larger reprocessing plant, together with waste storage facilities, at Rokkasho-mura.

Neither development has moved ahead as fast as was hoped for. And it was the realisation that these facilities were unlikely to be available in time to deal with the growing quantities of spent fuel being produced by the industry which led the Japanese electrical utilities, which run the power stations, to look elsewhere. Some used fuel from a small Magnox reactor built at Tokai-mura, north of Tokyo, in 1965 was already due to be sent to Sellafield. But the contracts signed with British Nuclear Fuels and COGEMA in the 1970s were of a different order. Roughly 80% of the spent fuel produced by Japanese commercial reactors is now due to be sent to Europe, over 5,000 tonnes in total. In return, an estimated 49 tonnes of plutonium could be returned to Japan from Europe by the end of the century.

Meanwhile, there have been further delays in the Japanese programme. The proposed reprocessing plant at Rokkasho-mura, until recently a remote fishing community, has aroused a particularly strong wave of anti-nuclear and environmental opposition. Construction work is not expected to be finished until the early years of the next century. Operation of the Monju fast breeder reactor has also been held up by the sort of technical difficulties which have dogged the European programme. At one time scheduled to start up in 1991, it has now slipped behind a revised 1993 start-up timetable.

The effect of these hold-ups is more than just irritating for the Japanese, however. Some of the plutonium to be delivered from Europe could be converted into MOX fuel, for which Japan intends to develop its own full scale fabrication plant. But one result of the delay in starting up the Monju fast breeder is that the delivery of plutonium aboard the *Akatsuki Maru*, transported at such cost and controversy, will be placed in storage, possibly for years. The net result of the contracts with BNFL and COGEMA is therefore that Japan is in the process of acquiring a surplus of plutonium for which it has no immediate use. This is something it has publicly stated it would never do.

Against the background of her post-war rejection of military expansionism, any stockpiling of potential bomb-making material by Japan has raised disquiet. But in the present political climate of the Northeast Asia region, the issue is particularly sensitive.

Despite international pressure, especially from the United States, the region is now in the grip of an anticipated escalation in nuclear weapons related production facilities. North Korea, which already has a small nuclear industry, is building a reprocessing plant capable of handling up to 300 tonnes of spent fuel per annum. It has already separated some plutonium, and is generally believed to be planning to build nuclear weapons. There is an ongoing dispute between North Korea and the IAEA about safeguards against military diversion of nuclear materials.

Its wary neighbour, South Korea, with a more developed nuclear industry, has made several abortive attempts to acquire reprocessing technology from France and Canada, but was blocked by the US. More recently, discussions have been held with BNFL, COGEMA and Russian nuclear plants about possible deals which could result in the return of plutonium from spent fuel. There is little doubt that the commitment of Japan to the plutonium economy has both added to tension in the region and encouraged the feeling among her neighbours that they should also follow suit. It hasn't helped that some right wing Japanese politicians have also spoken out against the long-standing ban on Japan's possession of nuclear weapons.

The much-publicised journey of the *Akatsuki Maru* at the end of 1992 opened up the controversy still further. In the United States it prompted renewed concern about reprocessing, and its direct effect on proliferation in the Korean peninsula. A Senate commissioned report by the General Accounting Office[25] pointed out, however, that although US policy is generally against reprocessing, this does not apply to Japan and Western Europe. Although the uranium fuel sent to THORP comes originally from the United States, the nuclear cooperation agreement between the US and Japan allows for separated plutonium to be returned for up to 30 years. The United States can only officially stop the shipments if they pose a threat to national security or significantly increase the risk of proliferation.

Some Congressmen have now called on President Clinton to put pressure on the British government to block the commissioning of THORP, a policy also pressed on John Major when he visited the US in early 1993. Clinton has said he will raise the issue with Britain, having looked at ways of combating proliferation. Senator John Glenn, one of the more outspoken US politicians on the issue, described the use of plutonium for fuel as a "high cost, low benefit, high risk exercise in futility".

An indication of the polarisation which has developed on the issue came with two adverts placed in the British press in June 1993, one from an organisation called Japanese Citizens Concerned About Plutonium, the other from ten Japanese electrical power companies. The latter, conceived by BNFL, urged the UK government to start up THORP as soon as possible; the former warned of economic disaster for the project if the Japanese commitment to THORP should falter.

Behind the scenes, there is growing uncertainty among some Japanese power companies that they actually want the plutonium economy their country publicly espouses, with all the difficulties and costs of both fast breeder and MOX programmes. Squeezed at the same time by anti-nuclear pressures at home and anti-proliferation pressure from abroad, the security of their contracts with THORP looks by no means as solid as it did when they first signed up in the heady days of the 1970s.

THE reason why Germany is sending such large quantities of its spent fuel to THORP for reprocessing is somewhat different from the Japanese motivation. In order to obtain an initial licence under the German "Atomic Act", power utilities must be able to show that they can deal with the aftermath of operating nuclear reactors for at least six years ahead. At present, the only acceptable solution for disposing of spent fuel under this arrangement is through reprocessing.

Until 1989, it had been planned to open a large reprocessing plant at Wackersdorf in Bavaria in order to reprocess much of the fuel from Germany's 20 reactors. Construction work actually began in 1985, but there were massive public protests on a scale unseen in Britain, including an occupation of the site by 100,000 demonstrators and pitched battles with the police.

The costs of building Wackersdorf also escalated, a familiar tale with nuclear projects, from DM 5.4 billion in 1985 to DM 7.7 billion four years later. Eventually, the German electrical utilities, concerned about the rising cost, the hazy future for plutonium-fed reactors and the political uncertainties, began to look elsewhere. They finally persuaded the government that they could use a combination of the already committed plants at La Hague and Sellafield instead.

In May 1989, the partly built Wackersdorf plant was abandoned, and is now ironically being used for recycling non-nuclear waste materials and manufacturing solar cells. Up to 2,500 jobs could be created, over 1,000 more than were promised by the reprocessing operation. Redevelopment of the site is being supported by the government, partly through compensation payments due from the electricity companies.

The German contracts with THORP, meanwhile, were welcomed by BNFL with open arms. The company even offered the Germans a lower unit price on the basis that the spent fuel was converted into MOX. There is also a clause which allows for their cancellation by the

Germans without penalty should some event outside their control, such as a change in national policy, occur. In the wake of the deal, a nuclear cooperation agreement was signed between Britain and Germany, hopefully setting the seal on a happy ongoing relationship. However, this was by no means the end of the story.

In contrast to Japan, Germany's nuclear industry has effectively been stalled from further power station construction by environmental pressures. The opposition Social Democratic Party (SPD) has specifically called for an end to reprocessing by German utilities, whether at home or overseas. These pressures have now been caught up in a major national energy debate, including proposals to change the Atomic Act.

The most important proposed change as far as THORP is concerned is that the German power companies would be allowed the option of direct disposal of their fuels instead of reprocessing. I shall return to the issue of this alternative system of dealing with spent fuel in more detail in Chapter 7. But the opportunity of a choice in waste management has at least been welcomed by the utilities. A 1992 report by one of the three largest companies, RWE-Energie, estimated that if nuclear power station operators could switch immediately to direct disposal they could save about DM 6 billion by 2005.[26]

The national debate has also threatened the future of the new MOX fuel fabrication plant at Hanau. Power utilities have warned that they will withdraw their support for the plant, being built by the Siemens company, unless there is agreement soon on a plutonium recycling policy. One possibility is that the national energy debate could lead to all use of plutonium fuels being jettisoned in favour of a consensus on continued nuclear power generation.

On current form, the German energy debate, now expanded to cover the whole spectrum of fuels, including fossil fuels and renewables, is likely to be protracted. The proposals for changing the Atomic Act have already been approved by the government, but have still to be ratified by parliament. But even without this change, there is growing doubt among German power companies, because of delays in developing the MOX fuel programme, about the whole plutonium economy.

The effects on THORP of these pressures could be dramatic. Without the German deliveries of spent fuel, some of which the

Germans have the right to cancel should unforeseen circumstances ("force majeure") intervene, the economic future for the project would look distinctly bleak.

ALTHOUGH Japan and Germany are the main overseas customers for THORP at present, accounting for over 50% of the "baseload" contracts in the plant's first ten years of operation, a number of other countries are also committed to send smaller quantities of spent fuel. These are Switzerland, Spain, Sweden, Canada, Italy and the Netherlands.

For most of these countries, agreement to the THORP contracts was taken at a time when reprocessing seemed more attractive and there was much less anti-nuclear sentiment at home. There are now de facto moratoria on further nuclear construction in Spain, Holland, Italy, Switzerland as well as Sweden, and none of these countries are likely to provide any future business. Only Switzerland has plans to develop any sort of plutonium economy, a decision prompted by the utilities themselves rather than national policy. Thirty nine members of the Swiss parliament have sent an open letter to the UK government registering their opposition to THORP.

It is therefore clear that, for a whole range of reasons, the overseas customers for THORP are far less committed to the project than in its early days. At the same time, given the uncertainties surrounding the whole future of reprocessing, there is no large pool of new business waiting to be ushered in. Quite apart from the issues already raised about security and proliferation in the transport of nuclear materials round the world, this could have a further basic effect—it could seriously damage THORP's prospects of making a profit.

6 The Economics of THORP

WHEN the THORP project was first conceived in the 1970s, the last thing to be considered was the possibility that it wouldn't make money. At the Windscale Inquiry the issue was scarcely debated. With only one other proposed plant (in France) in serious competition, offering reprocessing services to the world's nuclear industry was viewed as a seller's market. There also seemed every prospect that the industry would expand, bringing in fresh customers.

Almost two decades later, that rosy prognosis has radically altered. As a result of the developments already described in previous chapters, reprocessing is no longer a growing business. There is no market at all for reprocessed uranium, and an uncertain future for plutonium in mixed oxide fuel. Without the large domestic customers offered by the French nuclear industry, BNFL in particular has had to fight hard to achieve its present level of contracts. And the question of whether THORP will produce any surplus at all is now seriously under question.

What are the central features of a balance sheet for THORP in 1993? On the debit side, there are its construction cost, its operating expenses, and, most crucially, how much it will cost to dismantle. On the credit side, there is income from the customers, some of whom have already paid up front for the delivery of spent fuel. BNFL is a state-owned company, whose only shareholder is the government, but it is expected to operate like a commercial company, and generate a profit from its activities.

The cost of building THORP has risen steadily, from £300 million at the time of the Windscale Inquiry to £1.8 billion on completion in 1992. It was originally due to start operating in 1987. With the additional costs of associated facilities, including new waste treatment buildings, the total bill now reaches £2.8 billion. The plant is expected to have a working life of 25 years.

At the other end of the process, there is a substantial sum involved in dismantling THORP. This is because from the moment it starts operation, the building becomes contaminated with radioactivity. Many areas will be sealed up, and human entry made impossible during its lifetime and beyond. At the end of the day, the structure and all its mechanical parts will themselves have to be treated as radioactive waste. BNFL puts the cost of this "decommissioning" process for just the main THORP plant at £900 million.

As far as operating THORP is concerned, the costs have risen mainly because of the pressures on Sellafield to "clean up its act". From special treatment plant to reduce the radioactivity in sea discharges through to concrete vaults to house low level waste, BNFL has been forced to spend large sums on expensive environmental measures. One of the major elements in THORP's balance sheet will be dealing with the nuclear wastes generated as it goes along. Although, as indicated, some materials will be returned to foreign customers, over half the high level waste, and (if the substitution arrangements are agreed) all the low and intermediate level waste, will have to be dealt with at Sellafield. In 1990, the cost of disposing of just one cubic metre of intermediate level waste was set at £7,000.[27]

BNFL argues that, despite these heavy costs, and at least for the first ten years of THORP's operation, it has a profitable business. This is based on the series of contracts for reprocessing fuel agreed from the 1970s onwards (see chart on page 45). About two thirds of this business comes from the overseas customers already discussed. The other third comes from the two state-owned British nuclear power generating companies, Nuclear Electric and Scottish Nuclear.

Although early (pre-1976) contracts, which have no clause for return of wastes, would make a loss at today's prices, later contracts are extremely generous to BNFL, even allowing for additional costs to be passed on as they occur. This total business would almost fill THORP's expected annual throughput of 700 tonnes of spent fuel for a decade. Advance payments of £1.6 billion from customers have largely covered the construction costs. The net result, according to BNFL, is that the first ten years' income will not only cover all building, operating and future decommissioning costs, but will produce a profit of £500 million.

After the first ten years, the prospects look a lot less promising. Although on paper there are further contracts from both the

British nuclear companies and from Germany, these are by no means set in stone. We have already seen the pressures building up within the German power industry, partly because of proposed changes in the Atomic Act, which would allow an alternative to reprocessing, and partly because of the extra costs involved in using plutonium-based fuel.

In 1992, Greenpeace revealed details of a discussion document being considered by German Chancellor Kohl. This outlined a scenario in which no future contracts would be taken up with BNFL and all existing agreements cancelled, even if this meant paying compensation. Among those involved in drawing up the document were the two largest German power companies, RWE and Veba. If the Germans pulled out of just the second decade contracts because of a change in the Atomic Act, it is understood that they would suffer no financial penalties.

The threat of a German withdrawal has also cast a shadow over the British second period contracts. It would be uneconomic for BNFL to only reprocess British spent fuel at the already agreed price. But the companies themselves are not that keen either. In the case of Scottish Nuclear, which operates two AGR power stations, the company has already applied for permission to construct a dry storage building at its Torness site, halting its flow of fuel to Sellafield. The Nuclear Electric position is more complicated, but it has essentially only signed up for further processing of oxide fuel because it wanted an overall deal for all its fuel, including the bulkier and more trouble-some canisters from the older Magnox reactors.

It is worth recalling the words of Alan Johnson, Site Director at Sellafield, recorded giving a lecture to staff during a 1989 Channel 4 television documentary. "Reprocessing is not necessary," he said, with alarming honesty. "In fact one or two of our important customers would love to cancel their contracts at the drop of a hat. We won't let them of course. But there could be circumstances when we might just have to."

As far as new business is concerned, the problems already described are probably the biggest deterrent for potential overseas customers. Having seen the difficulties associated with the Japanese traffic, it seems unlikely that they or others will be looking to develop new contracts. In a situation where fast breeder reactors look an ever distant dream, fresh uranium is cheap, MOX fuel continues to be an expensive option, and with alternative methods of disposal looking increasingly attractive, there is simply not much incentive, excluding the military motive, to follow the reprocessing route. Even apparently eager customers such as South Korea, with whom BNFL has been negotiating, may be simply excluded by international pressure against proliferation.

According to some economic analysts, such as Frans Berkhout,[28] there are also a series of potential sources of cost escalation which could make even the slim profit projection on THORP's first decade—£50 million per annum—look fragile. These include a renewed demand for the installation of krypton gas removal equipment, increased storage space for high level waste not returned to overseas customers, a delay in building the NIREX deep repository (requiring more storage space for intermediate level waste), and an increase in decommissioning costs. It is equally possible that more plutonium could have to be stored at Sellafield because of delays in its return to customers.

Berkhout estimates the potential cost increases attributable to all these developments at between £710 and £1210 million. If BNFL held out any hope of attracting existing customers into further business, it would find it difficult to simply pass on these costs, as is technically allowed under most contracts. As a result, they could swiftly eat into THORP's potential profit.

Doubt has also been cast on the whole basis of the deal between Nuclear Electric and BNFL, which sets a value on the total reprocessing of all the Magnox and AGR fuel from English and Welsh nuclear power stations at a massive £14 billion. Since the deal is between two state-owned companies, both of which have effective monopolies in their respective fields, it has been questioned how realistic the costings are in such an arrangement. Is Nuclear Electric's income, a large proportion of which comes from a special levy of about 10% on all electricity bills, being used to subsidise THORP?

At the time of writing, the agreement between BNFL and Nuclear Electric has still not finally been signed (although negotiations began in 1986) because of a dispute about who will take liability for risks such as future changes in safety or environmental regulations. According to a report by Berkhout and William Walker,[29] British electricity consumers will pay £1.7 billion more than is necessary to have spent fuel reprocessed at THORP. The authors suggest that most of BNFL's profit will in practice come from the UK contracts.

Given these financial risks and liabilities, all of which would ultimately implode on the British Treasury (and the British taxpayer), how much would it cost to abandon THORP, as has been advocated by organisations like Friends of the Earth and Greenpeace, and simply not allow it to start up? According to a private assessment prepared for BNFL by financial consultants Touche Ross, the answer is about £900 million, although this figure is for "loss to the country" as a whole, not just the BNFL balance sheet.

Critics point out that this projection ignores the full range of cost escalations possible if the plant was allowed to open, making it more expensive to run than to mothball, and also doesn't consider the possibility of shifting existing customers towards a new dry storage option. Meanwhile, the full Touche Ross report has not been made public, despite many demands from environmental groups that this would be the only fair way to assess the calculations.

It now appears that, despite the company's confident public face, few parties to the enterprise would be totally distraught if some solution were produced which would disengage them from their contracts and not leave them carrying the can—in this case several thousand tonnes of spent nuclear fuel. The customers, especially the major clients in Japan and Germany, are fearful, however, that any unilateral withdrawal would result in their not only forfeiting their investment but having to take back their already delivered fuel. BNFL, for its part, knows that cancellation would make it liable for heavy compensation payments to "disappointed" customers, as well as still having to deal with the stockpile of fuel. The situation has been aptly portrayed as a nuclear "prisoners' dilemma", with all sides trapped in a game of blind man's bluff in which the impossibility of their collaboration means that the resulting decision will fail to satisfy any of them.[30]

One solution, offered by Greenpeace among others, is that the existing contracts for THORP should not be terminated, as BNFL assumes, but instead translated into new agreements involving dry storage of the spent fuel already delivered. On this analysis, the financial outcome would be more favourable to BNFL (and ultimately the government) than straight cancellation. From the customers' point of view, they would continue to receive a waste management service, but would not be burdened with either the costs or hazards of dealing

Pipework gallery in the Chemical Separation area

with unwanted waste resulting from reprocessing, nor with an excess of unusable and hazardous plutonium.

In practice, only the British government could begin to unravel this tangled web of bluff and counter-bluff and produce a solution which would both satisfy the overseas customers and leave BNFL with some secure future business. I shall return to the political background in the final chapter.

THERE is one other economic issue to be taken into account, and one less easy to quantify in straight monetary terms. This is the effect on the regional and national economy if a major project of this sort does not go ahead.

Whatever else, there is no denying the scale of the THORP investment. It has already employed 5,000 people over the ten year
construction period (8,000 on the whole development package).
BNFL says that when it is fully operational, THORP and its associated plant would directly employ over 2,000 people. A further 3,000
or so jobs would be "supported" by the scheme, including suppliers
and local services, most of those in West Cumbria. More jobs would
also be created by the proposed MOX fuel fabrication plant, which
would only be built once THORP had started operation. If THORP
was abandoned now, the company predicts that a minimum of 1,250
jobs would be lost.

In an area where unemployment is close to the national average,
BNFL dominates the local job market. It already employs over
10,000 people on the Sellafield site, including contract workers,
and claims to inject an annual £60 million into the Cumbrian
economy.

Some critics argue that Sellafield has effectively suffocated the
local economy, making it difficult for other enterprises to flourish.
Because of its role at the end of the nuclear fuel cycle, it has also
failed to respond to the present stagnation in the nuclear industry and
diversified, a policy being pursued by the UK Atomic Energy
Authority (now known as AEA Technology) with some success.

"BNFL rules the roost," says Martin Forewood of the main local
protest group CORE (Cumbrians Opposed to a Radioactive
Environment). "It has entrenched itself into every aspect of our
society, whether it's education or health or the arts. There is nothing
else." He believes that potential investors are actually put off by the
region's "Sellafield stigma", and that a halt to expansion might
encourage them in.

There is also disagreement about the figures themselves. BNFL
has blamed recent large-scale redundancies in the company on the
delay in starting up THORP. Others argue that lay-offs were
inevitable once construction of the project was complete, and point
to a 1986 House of Commons Environment Select Committee report
which recommended a study into how the construction workforce
could be re-employed elsewhere.

Although the projected employment losses from abandoning
THORP might seem large, they bear comparison with what has
happened in other energy-based companies. Nuclear Electric is in the

process of shedding 2,000 jobs, National Power about 10,000 and British Gas 20,000.

If the alternative policy of dry storage of nuclear fuel were pursued, there would also be new jobs created. One estimate is that a Sellafield dry store would involve 6,800 construction jobs, employ 500 operational staff and a further 500 in support employment. Even BNFL admits that about £300 million could be saved by building a dry store instead of operating THORP, assuming that there was no lost revenue from cancelled fuel contracts. Such a sum invested in energy efficiency, for example, could create 1,200 jobs for ten years, based on experience in the United States. At the same time, given the time-scales involved in nuclear waste management, Sellafield as an industrial site is in the comparatively privileged position of not looking forward to early retirement.

It is an extremely difficult task to assess the value of new jobs created against the scale of environmental damage which has been laid at Sellafield's door, but several points can be emphasised. Firstly, the number of jobs lost are comparatively small compared with other recent closures, such as in the coal mines. Secondly, as with all nuclear waste activities, there will be more work to come, especially if some ultimate disposal system were established at the site. Finally, if reprocessing is an unacceptable activity then it should be stopped in the common good, including for the benefit of those waiting in the job centre queue.

In the current economic climate, however, it is likely to be the balance sheet questions of whether THORP will actually make a profit which will carry the most weight with those who come to make the final decision. Even here, as we have seen, THORP is fundamentally weak.

7 The Dry Storage Option

IN 1992, a unique event took place in the recent history of nuclear energy in Britain. A public inquiry was held at Dunbar, East Lothian into a major nuclear power project at which none of the main environmental pressure groups made any objection.

The reason for this was simple. Even though they didn't necessarily agree with every aspect, the proposal by the Scottish Nuclear company to build a dry store for spent nuclear fuel was considered a much more preferable idea to its big brother on the Cumbrian coastline.

For the nuclear industry itself, the extended storage of spent nuclear fuel has emerged as an increasingly attractive option over the last 20 years. But just as the reprocessing route has lost much of its justification, so the power station operators are literally running out of storage space. Whilst some nuclear stations might expect operating lives of over 30 years, many do not have fuel storage capacity for more than fifteen.

At present, fuel rods removed from a reactor core are in most cases initially stored under water inside custom-built "flasks" at the power station until some of their shorter-lived radioactive elements have run their course. Instead of then being transported to a reprocessing plant, where they would continue to be "wet" stored until ready to be dissolved, they could alternatively be transferred to a dry storage plant. This would either be located on the site itself or at some central location where fuel would be delivered from several power stations.

Compared with THORP, a dry store is a relatively simple engineering project. Once dried, the uranium fuel rods are loaded remotely into either vertical or horizontal channels or containers, made either from concrete or steel. These storage containers are themselves sealed within a gaseous environment. The heat generated as the radioactivity within the fuel gradually decays is removed by the constant flow of cooling air through the vault.

Apart from being cheaper than reprocessing, dry storage avoids some of the disadvantages of a wet store. These include the need to chemically control and circulate the cooling water, much more intensive surveillance, and comparative inflexibility in terms of physical movement of the fuel. (Some nuclear utilities still prefer wet storage, and are working on improved pool designs.)

There could still be problems, including a breakdown in the vital heat removal system or gradual corrosion in the stored fuel. As far as nuclear waste is concerned, however, there would be a substantial reduction compared with reprocessing. Because there is no chemical dissolving with dry storage, the total volume of radioactive material would be increased only threefold by storage, as opposed to over 50 times by reprocessing.[31] Discharges from a dry store would be minimal compared with the outflow from THORP. Overall, the radiation dose to both workers and members of the public would be much less than with reprocessing. Waiting 50 years or more before doing anything further with the fuel would also reduce the radiation exposure to operators.

Different countries have come up with different versions of the dry store system. In Germany, several regional dry stores are already planned. The first, opened at Ahaus in 1992, has a capacity for 1,500 tonnes of spent fuel. This represents about three years' output from the country's power stations. In the United States and Canada, dry stores have been built next to individual power stations. There are a range of competing designs of storage container. One estimate is that by the end of the century, roughly 80% of spent fuel discharged from nuclear reactors around the world will be stored rather than reprocessed.[32]

As British Nuclear Fuels is eager to point out, dry storage of spent fuel does not present a final solution to the aftermath of nuclear power. Dry stored fuel could still be reprocessed, for example, even after many years. But an alternative "final solution" would be to dispose of the spent fuel, after a period of storage (at least 50 years), into an underground repository, a similar method to that proposed for the vitrified blocks of high level waste left over after reprocessing. A similarly robust network of chambers would have to be devised, and the siting of the repository would have to satisfy equally demanding criteria for protection against a range of intrusions, from subterranean water flows to earthquakes.

Reprocessing and the Dry Store Option

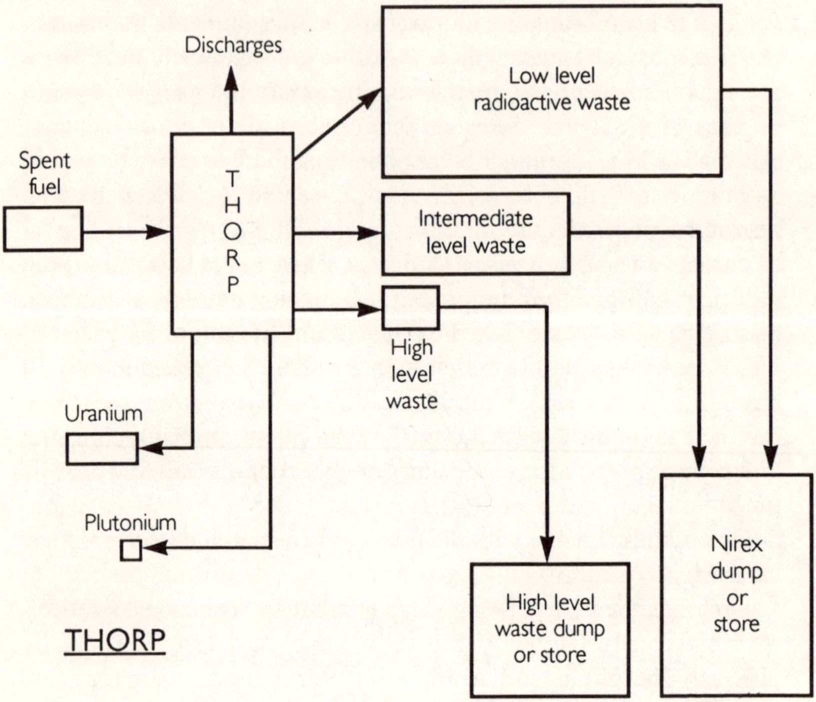

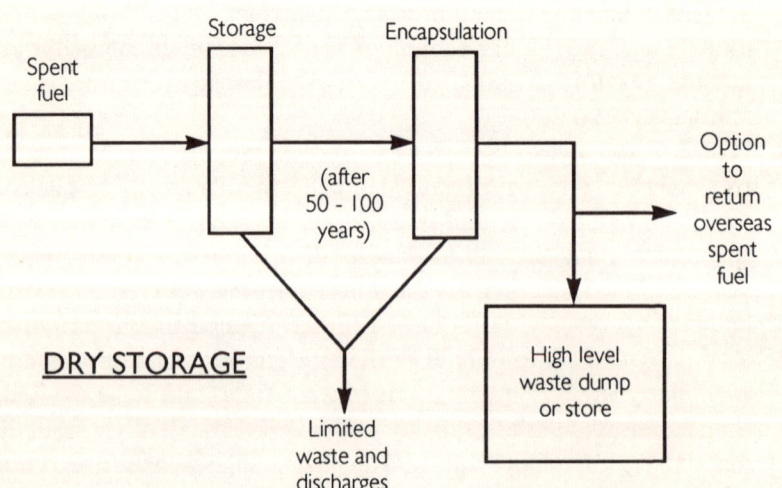

THORP

Spent fuel → THORP

Discharges

Low level radioactive waste

Intermediate level waste

High level waste

Uranium

Plutonium

High level waste dump or store

Nirex dump or store

DRY STORAGE

Spent fuel → Storage → (after 50 - 100 years) → Encapsulation

Limited waste and discharges

Option to return overseas spent fuel

High level waste dump or store

Technically described as "direct disposal" of spent fuel, this policy has been officially adopted in Canada, Finland and Spain, with the most advanced work being done in Sweden and the United States. In Sweden, the plan is to dismantle the fuel as though preparing it for reprocessing and then "encapsulate" it inside copper canisters. Some additional low and intermediate level waste would be produced by this process. The encapsulated fuel would then be sent to the underground repository.

From the industry's point of view, a major advantage of burying the spent fuel underground is that it limits the amount of ongoing management required. Others see it as a crude, "out of sight, out of mind" policy. In both Sweden and the United States, as with all versions of underground burial of highly radioactive material, there has been substantial opposition. To that extent, both reprocessing and storage present the same ultimate dilemma—what to do in the longer term.

One other suggested solution is to keep the material on the surface in a dry store for as long as is necessary, or until another, more acceptable idea comes along. This is what Greenpeace advocates. The major advantage of this is that the fuel can be carefully monitored and retrieved at any time. But there are concerns that above ground storage over long periods of time, effectively centuries, would present unacceptable technical and surveillance problems.

Even so, it is clear that an increasing number of nuclear countries are now choosing the dry store option as at least an interim measure until it is clear where the technology and science of ultimate disposal is leading.[33]

IN Britain, consideration of dry storage has been largely forestalled by the dominant view that reprocessing was the only way to treat spent fuel. Fuel elements from the early British Magnox power stations are much less easy to store over long periods in cooling ponds because they corrode. The former chairman of the Central Electricity Generating Board (CEGB), Lord Marshall, admitted in 1987 that his environmental critics were right, and it would have been better to use dry storage for Magnox fuel. "As a direct result of that failure of policy-making, we have had radioactive leaks at Sellafield and a loss of public confidence in that plant and in nuclear power generally," he wrote privately to the then Energy Secretary Peter Walker.[34] A dry store was in fact built in 1971, and still operates, for the last Magnox station in the series at Wylfa in North Wales, but was never repeated.

In the mid-1980s, the then Central Electricity Generating Board did consider building a central dry store for spent AGR fuel at Heysham in Lancashire. This plan was abandoned in the run-up to the privatisation of the electricity supply industry.

In 1990, however, the mould was finally broken when Scottish Nuclear announced that it was considering dry storage for most of the spent fuel produced by its two nuclear stations. The company was driven to this decision, it said, by the rising costs of reprocessing and its lack of need for recycled plutonium or uranium. British Nuclear Fuels still describes the decision curtly as "sending out the wrong messages. It's not helpful."

The Torness dry store is based on a design used at the Fort St. Vrain power station in Colorado, which opened in 1991. This has also been selected for the Paks power station in Hungary, and ironically uses the same basic system as the original Wylfa model. The dried fuel is held inside aluminium coated carbon steel storage tubes in a dry argon gas atmosphere. The tubes are sealed to keep them isolated from the cooling air, which flows through the store by natural convection. A similar store would be built at the other Scottish Nuclear site at Hunterston. They would both be designed to store fuel for up to 100 years after the final loading in the power stations.

Various comparisons have been made between the relative costs of dry storage and reprocessing. Scottish Nuclear says that it would save about £45 million per annum by operating dry stores at both Torness and Hunterston. This represents roughly 10% of the annual operating costs for the two stations. Scottish Nuclear's calculations include a figure for eventual disposal of the unreprocessed fuel, on the assumption that it would go to the same repository as Sellafield's high level waste. Internationally, the Nuclear Energy Agency, part of the Organisation for Economic Cooperation and Development, has estimated that reprocessing costs about twice as much as conditioning and direct disposal,[35] although a more recent NEA study shows the figures to be closer.

As far as THORP is concerned, would it be practicable to shift fuel from its present wet stores to a dry store? BNFL says that it would be technically difficult, especially in the case of British AGR fuel which has already been removed from its outer containers, and some of which has started to corrode. Other spent fuel from overseas stored at Sellafield might present fewer problems, as would fuel still to be delivered. BNFL could even consider providing dry storage in the country of origin as part of a revised contract. As was explained in the previous chapter, a switch to dry storage, even at this advanced stage of THORP's development, could have distinct economic advantages.

ABOUT BRITAIN'S NUCLEAR DUSTBIN

- They made
- They

DAILY Mirror

NEWSPAPER FOR THE NINETIES

THE KRYPTON FACTOR

SHOCK REPORT

New link between nuclear industry and child cancers

TOMB OF DOOM

A MASSIVE atomic tomb is being planned for the coast of England.

British Nuclear Fuels want to build the deadly dustbin at their Sellafield reprocessing plant in Cumbria.

More cancer claims pending against BNFL

New nuclear waste dump for Sellafield

8 The Politics of THORP

BRITISH Nuclear Fuels is a public company. It is therefore in theory open to the increasingly rigorous analysis accorded all state enterprises. But it also operates in what is seen as a strategic industry, one that not only involves a fuel source which has been consistently favoured by British governments, but has ramifications for the country's defence. This ambiguous status has given it an unusual degree of protection during a history which has been peppered with bitter criticism of its activities.

There have been fierce protests about the Sellafield site from well before the time it changed its name from the original Windscale in a vain attempt to expunge its negative associations. At the time of the Windscale Inquiry, the first major public rally against nuclear power in Britain was organised by Friends of the Earth. Since the emergence of the Greenpeace environmental campaign in the late 1970s there have been regular examples of "direct action", most notably in 1983 when underwater divers tried to block the discharge pipe, and discovered such extensive contamination that local beaches were closed.

British Nuclear Fuels tried to counter the wave of bad publicity by spending millions on its own public relations, including opening a £5 million visitors' centre at the site. The issue of THORP itself, however, remained relatively dormant until the late 1980s, when the proposed privatisation of the electricity industry opened up a debate about nuclear economics. The rising costs of reprocessing and waste management at Sellafield were a central part of that debate. As the argument developed into whether reprocessing was needed at all in the nuclear fuel cycle, THORP began to be seen as a potential financial liability.

During the summer of 1992, Greenpeace launched a renewed attack on THORP, highlighting the dangers of nuclear materials transport through a protest at Dover Docks, and the dangers of sea discharges by landing the Irish rock group U2 on the Sellafield beach. By this time the construction of THORP was nearing completion, and the building was due to start processing its first fuel at the beginning of 1993.

In order to start operating THORP, however, British Nuclear Fuels needed agreement for the increased discharges of radioactive material already described. The first government agency to put its head above the parapet was therefore Her Majesty's Inspectorate of Pollution (HMIP).

As a public company operating in a heavily regulated industry, BNFL has involvement with a number of government departments. Safety of its plant is generally overseen by the Nuclear Installations Inspectorate, part of the Health and Safety Executive. The Department of Health itself is also involved. Transport of nuclear fuels involves the Department of Transport. BNFL's general policy and investment decisions are supervised by the Department of Trade and Industry, whilst its finances involve the Treasury. Its military connections bring in the Ministry of Defence. HMIP, which is responsible to the Department of the Environment, itself works in conjunction with the Ministry of Agriculture over radioactive discharges.

Although not obliged to by law, the Pollution Inspectorate can allow a period of public consultation over new authorisations for radioactive discharges. This was originally due to start in August 1992. But a disagreement about the accuracy of some of the information supplied by BNFL led to it being delayed. In the end, the draft authorisations for discharges were not submitted for public consultation until November 1992. Even then, a further fortnight was added to the two month consultation period because of a further small inaccuracy in details of emissions from THORP's chimney.

The consultation process produced an unprecedented torrent of opposition. Encouraged by Greenpeace and others, thousands of individuals and organisations wrote in, not only objecting to the proposed discharges, but demanding that a full public inquiry be held. The Secretary of State for the Environment has the right to call an inquiry under the 1960 Radioactive Substances Act.

Altogether, over 64,000 objections were registered, the majority calling for a public inquiry. They included representations from 108 local authorities. Amongst these were detailed analyses as to why the discharges would be detrimental to people's health and to the

What the papers say

The need to conserve uranium has gone, the prospects for FBRs have diminished, yet THORP, the middle link in a chain which has been broken, will churn out uranium and plutonium for which there is no need. . . BNFL is about to start up a plant that even the nuclear industry agrees would not be built today.

Power in Europe, FINANCIAL TIMES, 28 August 1992

Now another large nuclear blunder is looming. The Thermal Oxide Reprocessing Plant (known as THORP) at Sellafield is ready to start operating. It should not be allowed to do so.

THE ECONOMIST, 19 December 1992

The contracts which Western utilities signed in the 1970s and 1980s to reprocess their nuclear waste have taken on a life of their own. But the environmental and economic cases for reprocessing have faded. Reprocessing does nothing to help manage waste from modern reactors: plutonium fuel is dearer than uranium, and likely to stay that way. It would be a far sounder policy to run down Western reprocessing.

THE ECONOMIST, 5 June 1993

As happened in the case of Concorde, no politician seems to have the far-sightedness to halt a programme which now has such a physical reality and would be so embarrassing to drop. These politicians will not in forthcoming decades pay the political price of mistakes made today. Yet the truth is that Thorp needs to be stopped as soon as possible.

THE INDEPENDENT, 29 June 1993

The need for plutonium fuel has receded. As Russia dismantles its nuclear arms, the world will be awash with uranium—enough to fuel all the nuclear power plants now planned or in operation for at least 50 years. President Bill Clinton would be prudent to put a global ban on production of fissile materials at the center of his nonproliferation policy.

NEW YORK TIMES, July 1993

Cumbrian environment. There were 19,000 responses in favour, many from Sellafield workers. 15,000 letters arrived on the very last day of the consultation period, lending weight to the story that a desperate final push was made by BNFL to boost its figures.

The level of objection to the discharge authorisations inevitably gave the government pause for thought. But this is by no means the only evidence of substantial public opposition to THORP. Its economics have been heavily undermined by a number of analysts. Much of the heavyweight press has moved steadily against the scheme, including the *Financial Times* and the *Economist*. And there has been a growing chorus of international opposition.

The Nordic countries in particular, including Norway, Iceland, Sweden and Denmark, have expressed concern about the increased level of radioactive emissions, especially of krypton. The historical effects of Sellafield discharges can already be traced right round the north of Scotland, across to Scandinavia, and even as far away as Greenland. In April 1993, the Swedish environment minister, Olof Johansson, wrote to his British counterpart on behalf of the Nordic Council pointing out that "if the discharges from the THORP plant lead to unacceptable pollution and interfere with the legitimate uses of the sea, such as fishing, the question of the closure of the plant has to be addressed".

The Irish government, whose coastline faces Sellafield, has been a consistent opponent of THORP. In repeated representations to Britain, especially over accidents at the site, it has stressed the additional risks to the health and safety of the Irish population posed by opening the new plant. Apart from opposing the whole development, it has called for substantial reductions in the proposed discharge levels.

Most recently, in June 1993, the Paris Commission, a European inter-governmental organisation concerned to reduce marine pollution, passed a resolution calling for no new discharges to be allowed from reprocessing plants without certain things happening. These were: a clear justification being made for the discharges, a full assessment made of their impact, and the "best available techniques" being used to reduce them. The resolution was agreed by a clear majority of nine countries, including Germany, with two "reservations" (France and Belgium) and only one vote, Britain's, firmly against. Although the Commission has no statutory powers, it was a considerable vote of no confidence in THORP. Failure to comply with its terms will set Britain yet again at odds with its European partners.

There is also the ongoing discomfort in some sections of the United States administration about the expansion of reprocessing in Europe. Two Bills are currently progressing through Congress whose clauses would further emphasise the dangers of proliferation resulting from any trade in plutonium.

WITH an increasing sense of frustration, British Nuclear Fuels has fought back hard against its critics. As other justifications for THORP have floundered, it has concentrated more and more on the straight economics. "The customers are prepared to send their fuel here for reprocessing and BNFL is prepared to reprocess it," I was told when I visited the plant. "We operate to make a profit. This will be a very successful commercial undertaking."

During early 1993, BNFL elicited a series of letters of commitment from its overseas customers, including the national newspaper advert from the Japanese electricity utilities. The company says it is losing £2 million every week because of the delay in opening the plant.

A surge of concern has been expressed by BNFL for both the workforce and the wider Cumbrian economy. Trade unions at Sellafield, led by the GMBU, have worked closely with the management in an attempt to salvage the hoped for jobs. Local newspapers warn of an "economic and employment catastrophe". The plant's GMBU branch secretary told me that if THORP didn't open, "you might as well shut the whole place down and walk away." Both the local Copeland District Council and Cumbria County Council have swung firmly behind THORP because of the jobs involved.

BNFL has also successfully maintained a number of important political friends at Westminster, and not just on the government side. The local MP to Sellafield, Jack Cunningham, is also Labour's spokesman on foreign affairs, and a long-standing and influential member of the opposition front bench team. Together with other pro-nuclear MPs, such as Dale Campbell-Savours and Tam Dalyell, he has forcefully argued the BNFL case. Both Cunningham and John Smith, the Labour leader, are sponsored by the GMBU, the main trade union at Sellafield.

Torn between its pro- and anti-nuclear wings, official Labour

Party policy on THORP is one of "waiting to be convinced" sceptical neutrality. Through its environment spokesman Chris Smith, Labour is currently calling for a full Environmental Impact Assessment of the scheme, as well as demanding the release of various documents, including the Touche Ross report on the plant's economics and details of the customer contracts (especially cancellation clauses). Only when this is all on the table does the party consider a public inquiry relevant.

But the effect of Labour's "considered approach", despite its implied criticism of reprocessing, has still been to blunt parliamentary opposition to THORP, leaving the main criticism to come from the Liberal Democrat benches. The Liberals have long opposed nuclear power, including reprocessing. David Steel and Russell Johnston, both still Scottish MPs, spoke out against the plant in the original 1978 debate. There has also been opposition to THORP from some Northern Ireland members.

BNFL has been in an equally good position to lobby behind the scenes in Whitehall. In July 1992 its previous chairman, Christopher Harding, resigned in favour of John Guinness. Until then, Guinness was a senior civil servant at the Department of Energy, which was incorporated into the DTI after the 1992 general election.

IN the aftermath of the public consultation process, meanwhile, which finally ended in January 1993, the government found itself in an uncomfortable dilemma. On the one hand the "authorising departments", HMIP and MAFF, had by now reported that as far as the new discharge applications were concerned, they had not been persuaded to change their minds. Apart from some minor amendments, BNFL could have its approval. On the other, a large number of fundamental objections had been raised to the scheme. Many of these reflected the logical point that in so many areas, from plutonium demand through to radiation risks, the ground had shifted substantially since Lord Justice Parker sat in Whitehaven town hall.

It had also become increasingly clear that the economics of THORP looked decidedly thin. Not only was the profit margin predicted by BNFL in the first ten years nothing to write home about, but there was a growing checklist of uncertainties, not least the threatened German pull-out, which could quickly wipe it out.

Treasury officials even held a meeting with representatives from Greenpeace to hear their financial prognosis.

At the same time, Britain was beginning to look isolated over THORP in both the European and international arenas, especially after the Paris Commission decision. The British Euro-MP Llew Smith also nearly succeeded in persuading the European Parliament in July 1993 to pass a series of resolutions severely criticising the international traffic in spent fuel.

Finally, there was the threat of a legal challenge if the government failed to call a full public inquiry into the proposed THORP discharges. Legal advisors told Greenpeace that such a challenge stood a strong chance of succeeding if no further response was made to the criticisms raised in the 60,000 letters to the Pollution Inspectorate. At one point the Attorney General was consulted to discover

The THORP plant viewed from the Management Centre

how the government could avoid a "judicial review" of any decision it made.

In the end it took until June 28 for the newly appointed Secretary of State for the Environment, John Gummer, to announce in the House of Commons that there would be a new, second round of public consultation about THORP, this time looking at the whole question of its justification. Using the occasion of a Liberal-inspired debate, the announcement said that the new consultation would last for two months, ending on October 4, and would be based on three new reports from the government and BNFL, as well as an analysis of the first consultation process by HMIP and MAFF.

At the same time the government firmly opposed the motion from the Liberal Democrats, which said there were "increasingly strong economic, environmental and proliferation reasons" for not opening THORP, and called for a fresh public examination of the arguments, including full disclosure of economic information such as details of reprocessing contracts. The parliamentary debate, the first about THORP since 1978, lasted less than three hours. With Chris Smith voicing criticism from Labour's front bench, but with the party as a whole abstaining, there was a resounding majority for a government amendment supporting "the commissioning of the plant at the earliest practicable date". Even after that, the government still announced, in launching the second consultation period on August 4, that it had not ruled out the possibility of eventually calling a public inquiry.

The illogicality of the government's position was quickly seized on. The "Statement of Government Policy" on reprocessing and THORP almost precisely mirrored the British Nuclear Fuels view. All the background documentation provided for the public consultation amounted to a reiteration of BNFL's previous position on the plant. There was no independent analysis. The general public, or interested parties, were expected to respond to this in a vacuum. And yet the government was implying that it still had an open enough mind on the issue to be considering a full public inquiry. To most observers it looked like a rubber stamp wrapped up in a desperate attempt to avoid a challenge in the courts.

In fact, this apparent contradiction—and the lengthy period it took for the government to respond to the original protests—reflects a much less uniform view on THORP behind the scenes than the parliamentary bravado suggested. Although the Department of Trade, whose Energy Minister, Tim Eggar, defended BNFL in the June debate, is extremely keen on THORP, especially because of the overseas trade involved, other departments are much more sceptical. The Treasury in particular, which has been keeping a tighter and tighter rein on government expenditure, has been worried that it might end up with a liability. The Foreign Office, not directly involved with THORP, is reported to be concerned about the international trade in plutonium. In the middle, the Environment Department, now headed by an admitted nuclear enthusiast, John Gummer, is uncertain which way to jump. At one stage the whole issue was submitted to a special Cabinet Committee to see whether some agreed path could be found through the conflicting views.

There is other evidence of shifts in the Whitehall position. Although government policy is officially in favour of THORP, there has been an important change in its attitude towards reprocessing as a whole. At the 1992-3 Torness public inquiry it was made clear that the government now has no clear preference between the alternative approaches of reprocessing and dry storage. A Scottish Office civil servant, speaking for the government, told the inquiry that the economic advantages of obtaining uranium from spent fuel had changed, and that "the reprocessing route did not appear to offer any immediate and significant advantages from a waste management point of view".[36] In civil service jargon, this was a way of saying that there has been a basic shift in policy.

More dramatically, there is a fascinating passage in the report produced in June 1993 for the US Senate on plutonium shipments to Japan. Having met various British officials during its investigation, the report notes that "one British government official told us that the rationale for operating THORP is no longer valid because THORP cannot be a financially successful venture. He further contended that without economic justification to engage in commercial reprocessing, the basis for reprocessing in the United Kingdom has collapsed."[37]

The irony of the present position is that whilst some Whitehall mandarins, and even some members of the government, would be happy to adopt a position not too dissimilar to that of Greenpeace, their public face doesn't allow them to seriously consider the possibility.

Conclusion

"TO have THORP standing, complete, waiting for the start button to be pressed, week after week, month after month, is like finishing the Channel Tunnel but refusing to allow trains to run." The speaker was Neville Chamberlain, chief executive of British Nuclear Fuels, in July 1993. Setting aside the fact that pushing the button for THORP would immediately incur large decommissioning costs, as well as the Chunnel's own doubtful economics, his frustration is understandable.

But it is equally true that if anybody, including BNFL, were given a choice about whether to start building a large oxide reprocessing plant now, the answer would almost certainly be in the negative. THORP may be a marvel of modern engineering, but it is a marvel designed in a period of unreserved optimism about nuclear power that has now passed.

How then do the justifications for starting THORP stand up to scrutiny in 1993, when set against its disadvantages?

Firstly, it's argued that THORP will produce useful by-products, specifically uranium for re-use in existing reactors, and plutonium either as fuel for future fast breeder power stations or in a mixed oxide mixture.

It is clear that the uranium argument has disintegrated. BNFL scarcely mentions it any more. With plenty of cheap uranium on the market, stockpiles of reprocessed fuel, and no sign of a major upturn in the international nuclear industry, salvaging it from reprocessing spent fuel simply isn't economically justified.

The only other argument put forward by BNFL is that re-using uranium from spent fuel avoids the waste generated by spoil heaps at uranium mines. Apart from the fact that extraction is anyway much reduced at present, this is more a reason for cleaning up the mining process than one in favour of reprocessing.

The plutonium argument is more complicated. Whilst the fast breeder programme has all but spluttered to a standstill, there is now the prospect that plutonium could be used instead in mixed oxide (MOX) fuel. Again, this doesn't stand up to detailed scrutiny. MOX has only emerged as an argument for THORP in the last few years. It is more expensive than straight uranium fuel, fabrication plants have been plagued with problems, and it's clear that many countries wouldn't be interested in it if they had not already signed up for reprocessing.

At the same time there is tremendous pressure to do something with the plutonium, both practically because storage is expensive and the material degrades, but also politically because without any recycling of materials, the justification for reprocessing falls apart. MOX is being promoted as that justification.

Japan (THORP's biggest overseas customer) might seem a special case, partly because it still has pretentions to getting a fast breeder up and running. It also plans to build its own reprocessing plant. The reality is that neither project is yet operating, there are growing uncertainties among Japanese power companies, and the plutonium already returned from France has turned out to be a political and diplomatic nightmare.

Even if there were some realistic uses for plutonium, there are other strong arguments as to why they shouldn't be taken up. Quite apart from its hazards as a radioactive element, requiring immense caution whenever it is handled, it is one of the prime sources of material for the manufacture of nuclear bombs. It can be argued that those who want to will find a way, and that THORP will be operated within strict guidelines. But producing many tonnes of the material in separated form, and then transporting it all over the world, is just asking for the sort of breaches in international security which have already occurred. As already explained, there is already much more plutonium in store than there is any use for: THORP will greatly increase that inventory.

Secondly, it's argued that reprocessing at THORP is the best way to handle spent fuel from the point of view of management of the radioactive wastes.

This is an argument which clearly predates THORP, and on which we have some existing evidence from operations at Sellafield.

The truth is that although reprocessing removes some radioactivity in the uranium and plutonium, it leaves behind a much larger volume of hazardous material to be dealt with.

Although less radioactive, the low and intermediate level wastes will still have to be safely contained for long periods covering many generations. As steadily higher environmental standards have been demanded, this has proved an extremely difficult task. The current industry plan is to bury them in an 800 metre deep repository whose construction will require the removal of as much rock as came out of the Channel Tunnel. Like numerous other plans for the burial of nuclear waste all over the world, this has yet to receive approval even from within the scientific community. As far as the highly radioactive waste is concerned, there are no firm long-term plans for where it might be kept. Why, it must be asked, go on producing large volumes of waste material for which there is no clear disposal plan?

The alternative to reprocessing is to store the fuel, probably in dry conditions, a method now adopted by many countries and recently chosen by Scottish Nuclear. This produces very little extra waste material and allows a period of up to 100 years for the radioactivity to decay further before the fuel is interfered with. As recycling of spent fuel into uranium or plutonium becomes increasingly less attractive, this is likely to become the international norm.

One possibility is that dry stored fuel could eventually be disposed of underground. This is the Swedish plan, for example. Although the amount of material to be buried would be larger in volume than the highly active waste from reprocessing, they would both take up a similar amount of space in any repository. This is because both spent fuel and highly active waste continue to generate heat and cannot be packed tightly. The same criticisms can of course be made of direct disposal as disposal of highly active waste—that there is no agreed operating site anywhere in the world. But the problem with direct disposal of spent fuel at least does not also entail large quantities of other wastes.

There is one other important disadvantage of waste management at THORP. This is the fact that, as one of only two similar plants in the world, it involves a major international traffic in radioactive materials. This is not only potentially dangerous in terms of accidents, but it also breaches the general principle that countries should move towards self-sufficiency in dealing with their waste products.

As far as the customers are concerned, it certainly seemed, when many of them signed their contracts, that THORP offered a useful way to get rid of their wastes. In 1993, in a changed world of nuclear stagnation, with dry storage an attractive alternative, and with the prospect of the return of a whole range of materials for which they may have no use, the picture looks quite different.

Thirdly, it's argued that THORP will be safe and will have no detrimental effects on the environment.

THORP is a new plant, built to higher standards than previous Sellafield efforts. Greater use of remote handling should reduce radiation doses to the operators, for example. But reprocessing is still one of the messiest and most hazardous activities in the nuclear fuel cycle, and the record at the Cumbrian site has not been good. Even since the much-publicised 1983 Greenpeace incident, for example, there have been further accidents, notably the 1992 leaking of plutonium nitrate, which classified as a rare example of a Level 3 "serious incident" on the international scale. During the first half of 1993, there were over 20 reported incidents at the site.

As far as the general public is concerned, there is real worry that the operation of THORP and its associated plant will expose them to greater levels of radiation without any clear benefit. Greenpeace projections set the number of deaths from cancer worldwide from every year of the plant's operation at over 60. Friends of the Earth emphasise the risk to people in the surrounding communities, already living under the shadow of an unexplained leukaemia cluster. All this must be set against the background of a continuing debate about the alleged failure of international regulations to take account of the known effects of radiation.

Fourthly, it's argued that THORP will make money for UK PLC and create jobs in West Cumbria.

This has now become the main justification for THORP, particularly the income its operation will earn from abroad to help the UK balance of payments. In practice, even BNFL only has current expectations of making a profit from THORP in its first decade, out of a potential lifetime of 25 years. Even here, uncertainties in the costings and potential increases in such items as storage charges could easily overhaul the slender margins. The nuclear industry, including

Sellafield, has become notorious for cost over-runs and sudden dramatic increases in charges. It seems unlikely that all these extra costs could be simply piled on to the customers' bills.

It's also clear that, despite their public protestations to the contrary, many of the customers would be happy to abandon their contracts and come to some arrangement for handling fuel already stored at Sellafield or irrevocably committed. The main foreign customers, Japan and Germany, have their own peculiar reasons either for slowing down their input to THORP or, in the case of Germany, switching completely to a different spent fuel management system. Nuclear Electric is now looking over its shoulder at Scottish Nuclear's dry store enterprise and wondering whether it could travel the same route. Remember the prophetic words of Sellafield Site Director Alan Johnson in 1989, quoted on page 56 above.

Against that background, the letters of commitment to THORP which the company has elicited in recent months must be seen as yet another bargaining counter in the poker game of potential compensation payments. As far as BNFL is concerned, it is in a relatively strong position to negotiate a new deal, possibly based on dry storage of delivered or committed fuel.

On the employment front, THORP would undoubtedly provide jobs, although the numbers who will be disappointed must be set alongside the much larger figures for redundancies of long-standing employees in other energy-related industries. In some ways, the area has already had the best of the project over the boom time of the construction period. It is up to British Nuclear Fuels, which still has substantial cash flow from its other activities, to devise some alternative employment if the plant were not to open. As already emphasised, the Sellafield site as a whole is not about to go out of business.

IT could be argued that, if THORP has so many disadvantages, why hasn't it self-destructed years ago? Many of the arguments we have heard—about plutonium proliferation, unreliable economics, radioactive pollution, and the advantages of dry storage—were cogently presented by opponents at the original Windscale Inquiry. One explanation is that it has now developed its own internal momentum. It keeps going because there is now simply no energy to try and disentangle the commercial and political decision-making which has brought it so far.

Like any development with environmental ramifications, however, it's not right to judge it simply on commercial grounds. Even if THORP could make money, its social and environmental disadvantages must be set firmly on the disbenefits side of the scales.

Finally, it must be remembered that THORP is deeply unpopular across a wide range of environmental groups, from Greenpeace to the Town and Country Planning Association. On this occasion, the parliamentary consensus does not accurately reflect the public mood.

It's argued from inside that abandoning THORP would deal a fatal blow to nuclear power at a time when the industry is already under pressure because of its unfavourable economics. It would certainly set the seal on a particular vision of nuclear expansion. But it would be a body blow not a knockout. Nuclear power generation could continue without reprocessing, as other countries have shown.

Looked at more positively, there is the real possibility that the contracts for THORP could be transformed into fuel storage agreements, starting BNFL off in a new role as an expert in what is likely to become a growing area of nuclear fuel cycle management. If a new dry storage plant was commissioned, deciding not to open THORP would by no means be the end of the story.

As far as the structure itself is concerned, if THORP was abandoned before being irrevocably contaminated by radioactive debris, it should not be beyond the wit of Britain's industrious engineers, inventors and recycling experts to devise some ongoing use for the structure. There is a precedent for this. The Wackersdorf plant, on which about £1 billion had been spent, was also up and ready to operate in 1989. Now it is a non-nuclear recycling factory. At the worst, THORP could become a monument to the great expectations with which nuclear power launched itself in the post-war decades—the ultimate nuclear folly.

Further Information

British Nuclear Fuels,
Risley, Warrington, Cheshire, WA3 6AS
(0925 832000)

Cumbrians Opposed to a Radioactive Environment (CORE),
98 Church Street, Barrow-in-Furness, Cumbria, LA14 2HT
(0229 833851)

Department of the Environment,
2 Marsham Street, London SW1P 3EB
(071 276 3000)

Friends of the Earth,
26-28 Underwood Street, London N1 7JQ
(071 490 1555)

Greenpeace,
Canonbury Villas, London N1 2PN
(071 354 5100)

Nuclear Free Local Authorities,
(National Steering Committee),
Manchester City Council, Town Hall, Manchester M60 2LA
(061 234 3324)

References

1. Professor David Henderson, "Two British errors: their probable size and some possible lessons", Oxford Economic Papers, July 1977.
2. The *New Scientist* reports are collected together in *100 Days of History: The Windscale Inquiry*, Ian Breach, IPC, 1978.
3. "The perils of plenitude", *Energy Economist*, Sept 1991.
4. Cited in speech by William Dircks, Deputy Director General of the International Atomic Energy Agency, to the Atomic Energy Forum of Japan, April 1992.
5. "Comparison of the Radioactive Waste Arisings Generated by Reprocessing, Encapsulation and Storage of LWR and AGR Irradiated Fuels", Large & Associates, 1992.
6. "The Way Forward", UK NIREX (Nuclear Industry Radioactive Waste Executive), Nov. 1987.
7. "Windscale: The management of safety", Health and Safety Executive, 1981.
8. "Collective Doses to Europe from Sellafield Discharges", David Sumner, Greenpeace, 1993.
9. John Dunster, Chief Health Physicist of the UKAEA, speaking at the 2nd UN Conference on the Peaceful Uses of the Atom, Geneva, 1958.
10. "The Projected Environmental Impact of THORP", P.J.Taylor, in "THORP: An In Depth Investigation", Cumbrians Opposed to a Radioactive Environment, 1990.
11. *Windscale—the nuclear laundry*, Yorkshire Television, directed by James Cutler, first broadcast 1 Nov 1983.
12. "Investigation of the possible increased incidence of leukaemia in young people near Dounreay", COMARE, 1988.
13. "Results of a case-control study of leukaemia and lymphoma among young people near Sellafield", M.J. Gardner and others, in the *British Medical Journal*, 17 Feb 1990.
14. "The Safety of Nuclear Fuel Reprocessing at Windscale", G.R.Thompson, PERG, 1977.
15. "Sellafield: The Contaminated Legacy", Friends of the Earth, 1993.
16. See "Comments on Draft Authorisations by HM Pollution Inspectorate for Disposal of Gaseous and Liquid Waste from Sellafield", David Sumner, Greenpeace UK, Nov.1992 and "How Safe Is Safe?", Greenpeace UK, Sept. 1993.
17. "Climatic Aspects of Radioactive Trace Gases, in particular Krypton-85", R.Kollert and M.Butzin, study carried out for the German Bundestag, Kollert & Donderer, Bremen, 1989.
18. Dr.R.W.Seldon, nuclear weapon designer at the Lawrence Livermore Laboratory, cited in David Albright, "Can civilian plutonium be used in nuclear explosives? A review of statements by nuclear weapons experts", Federation of American Scientists, 1984.
19. Victor Gilinsky, Plutonium Proliferation and Policy, US Nuclear

Regulatory Commission, speaking in 1976, cited in Albright, as above.

20. See "Britain and Plutonium", William Walker & Frans Berkhout, International Security Information Service (ISIS) Briefing 31, 1992.

21. "British civil plutonium: production and fate", letter from K.Barnham, D.Hart, J.Nelson and R.Stevens, in *Nature*, 23 June 1988.

22. "Plutonium and Reactor Transmutation", A.G.Elayi, Radioactive Waste Management and the Nuclear Fuel Cycle, Vol.14(4), 1990, cited in "The Mox Myth", Friends of the Earth, 1992.

23. Nine previous shipments of small quantities of plutonium were recently revealed to have taken place from Europe to Japan between 1975 and 1992, at least five from Sellafield (*The Observer*, 9 May 1993).

24. Detailed reports on the hazards of reprocessing transport include: "Import/Export of irradiated fuel and Radioactive Waste to and from the United Kingdom", Large and Associates, July 1990; "The International Transport of Plutonium, Spent Nuclear Fuel and High Level Radioactive Waste: An Assessment of Safety, Security and Proliferation Issues", Nuclear Free Local Authorities, May 1991; "The International Transport of Civil Plutonium", NFLA, July 1993.

25. "Nuclear Nonproliferation: Japan's Shipment of Plutonium Raises Concerns about Reprocessing", US General Accounting Office, June 1993.

26. "Nuclear waste disposal through contracts with Cogema and BNFL", H-J.Dibbert and E.Passig, RWE-Energie, 1992.

27. "Fuel reprocessing at THORP: Profitability and Public Liabilities", Frans Berkhout, Centre for Energy and Environmental Studies, Princeton University, 1993, p.14.

28. See ref. 18 above.

29. "THORP and the Economics of Reprocessing", Frans Berkhout & William Walker, Science Policy Research Unit, University of Sussex, 1990.

30. See "Nuclear scheme that became a 16-year Whitehall nightmare", in the *Financial Times*, 8 May 1993.

31. See ref. 5 above.

32. *World Inventory of Plutonium and Highly Enriched Uranium 1992*, Albright, Berkhout and Walker, OUP/SIPRI, 1993.

33. According to evidence from Scottish Nuclear at the Torness dry store public inquiry (see ref. 35 below), the following countries are "moving to long term storage while retaining the option of direct disposal or reprocessing": Mexico, Belgium, Germany, Italy, Korea, India, South Africa, Taiwan, the former Soviet Union and most former Eastern Bloc countries.

34. Correspondence revealed in "Edge of Darkness", in *Time Out*, 23 May 1989.

35. Quoted in the Radioactive Waste Management Advisory Committee (RWMAC) 11th Annual Report, 1990.

36. "Report of Public Local Inquiry into Objections to Proposed Spent Fuel Store at Torness Power Station", 1993, para. 3.10.

37. See ref. 10, p.10.

The World at Your Keyboard
An alternative guide to global
computer networking

BURKHARD LUBER

Thousands of people around the world use their computers
to communicate with other people, often thousands of
miles away, who share their interests—whether related to
work or hobbies, or to political campaigning.

Many more could do the same, given a few words of
advice and the purchase of a modem.

This jargon-free book contains all the information a
computer owner needs to gain access to this alternative
world of open communication. It includes simple advice on
how to use the networks, in particular the nodes of the APC
(including PeaceNet in the USA and GreenNet in the UK),
send and receive electronic mail, and participate in the
numerous 'conferences' on environmental, peace, development and other political issues.

The author takes illustrations from his own work in peace
research. An up-to-date list of 'bulletin boards' in Russia
gives readers direct access to people and information
there.

Burkhard Luber is director of the Threshold Foundation in
Bremen, and lectures throughout Eastern Europe and the
USA on disarmament and peace research. He is editor of
the quarterly journal *Transcontinental Peace Newsletter.*

£7.99 paperback 160pp ISBN 1 897766 00 9

Chernobyl: The forbidden truth

ALLA YAROSHINSKAYA

Foreword by Professor John Gofman

An extraordinary account by a Ukrainian dissident journalist of the corruption and dishonesty that precipitated the world's worst nuclear accident to date, and tried to conceal its terrible outcome from the rest of the world as well as from the very people most affected.

From the date of the accident in 1986, Alla Yaroshinskaya fought the bureaucracy and the official lies to establish the truth of what happened, and why the cover-up took place. She was on a list for internment as a troublemaker, if the coup of 1992 had succeeded. Her family lives near Chernobyl, and she worked on a newspaper there. The story is told partly in the words of the people who suffered—and still suffer—from official neglect and deceit, and draws also on the lies of local and national functionaries and the advocates of the nuclear lobby, and the opinions of medical investigators.

Even after Boris Yeltsin officially abolished the Communist Party, resistance at all levels to exposing the truth was as strong as ever, while at local level the fight against glasnost took pride of place over the needs and interests of sick people.

The revelations are many and startling: how contaminated meat was shipped to non-contaminated areas of the Soviet Union to spread the radiation load; how top officials and government leaders deliberately lied and misled their own people, the press, foreign governments and the IAEA.

It's a sordid and tragic, yet moving story, with bitter lessons for any country with a nuclear industry.

Illustrated with evocative photographs by M. Metzel.

Alla Yaroshinskaya was awarded the Right Livelihood Award (the 'alternative Nobel prize') in 1992 for work in exposing the truth about Chernobyl. She now lives and works in Moscow.

£8.99 paperback 160pp ISBN 1 897766 03 3
Available November 1993